Editorial Policy

for the publication of monographs

In what follows all references to monographs are applicable also to multiauthorship volumes such as seminar notes.

§ 1. Lecture Notes aim to report new developments - quickly, informally, and at a high level. Monograph manuscripts should be reasonably self-contained and rounded off. Thus they may, and often will, present not only results of the author but also related work by other people. Furthermore, the manuscripts should provide sufficient motivation, examples, and applications. This clearly distinguishes Lecture Notes manuscripts from journal articles which normally are very concise. Articles intended for a journal but too long to be accepted by most journals usually do not have this "lecture notes" character. For similar reasons it is unusual for Ph.D. theses to be accepted for the Lecture Notes series.

§ 2. Manuscripts or plans for Lecture Notes volumes should be submitted (preferably in duplicate) either to one of the series editors or to Springer-Verlag, New York. These proposals are then refereed. A final decision concerning publication can only be made on the basis of the complete manuscript, but a preliminary decision can often be based on partial information: a fairly detailed outline describing the planned contents of each chapter, and an indication of the estimated length, a bibliography, and one or two sample chapters - or a first draft of the manuscript. The editors will try to make the preliminary decision as definite as they can on the basis of the available information.

§ 3. Final manuscripts should be in English. They should contain at least 100 pages of scientific text and should include

- a table of contents;
- an informative introduction, perhaps with some historical remarks: it should be accessible to a reader not particularly familiar with the topic treated;
- a subject index: as a rule this is genuinely helpful for the reader.

Lecture Notes in Statistics 97

Edited by S. Fienberg, J. Gani, K. Krickeberg, I. Olkin, and N. Wermuth

R. Szekli

Stochastic Ordering and Dependence in Applied Probability

Springer-Verlag
New York Berlin Heidelberg London Paris
Tokyo Hong Kong Barcelona Budapest

R. Szekli
Mathematical Institute
University of Wroclaw
Pl. Grunwaldzki 2/4
50-384 Wroclaw
Poland

Library of Congress Cataloging-in-Publication Data Available
Printed on acid-free paper.

Softcover reprint of the hardcover 1st edition 1995

Camera ready copy provided by the editor.

9 8 7 6 5 4 3 2 1

ISBN-13: 978-0-387-94450-0 e-ISBN-13: 978-1-4612-2528-7
DOI: 10.1007/978-1-4612-2528-7

PREFACE

This book is an introductionary course in stochastic ordering and dependence in the field of applied probability for readers with some background in mathematics. It is based on lectures and seminars I have been giving for students at Mathematical Institute of Wroclaw University, and on a graduate course at Industrial Engineering Department of Texas A&M University, College Station, and addressed to a reader willing to use for example Lebesgue measure, conditional expectations with respect to sigma fields, martingales, or compensators as a common language in this field.

In Chapter 1 a selection of one dimensional orderings is presented together with applications in the theory of queues, some parts of this selection are based on the recent literature (not older than five years).

In Chapter 2 the material is centered around the strong stochastic ordering in many dimensional spaces and functional spaces. Necessary facts about conditioning, Markov processes and point processes are introduced together with some classical results such as the product formula and Poissonian departure theorem for Jackson networks, or monotonicity results for some renewal processes, then results on stochastic ordering of networks, replacement policies and single server queues connected with Markov renewal processes are given.

Chapter 3 is devoted to dependence and relations between dependence and ordering, exemplified by results on queueing networks and point processes among others.

From the technical point of view the unifying thought of this course is to construct random elements on a common probability space to derive required properties almost surely. A unifying language is that of theory of point processes.

The main parts of the manuscript were prepared during my stay in Hamburg at the Institute for Mathematical Stochastics under the Alexander von Humboldt Fellowship in 1991 and in College Station at the Industrial Engineering Department of Texas A&M University, in 1992 under the Kosciuszko Fellowship. Special thanks are directed to professor Hans Daduna from Hamburg and professor Ralph Disney from College Station for their hospitality which made my work a pleasure. Thanks also go to professor Robert Foley from Georgia Tech for reading some parts of the manuscript and giving helpful remarks.

Contents

Chapter 1

Univariate Ordering

Mathematical modeling in engineering and other fields which use as a tool the theory of probability involve concepts of comparison of random variables. The usual concept of *partial order* is applied to random variables, random vectors, counting processes etc. Random behavior of variables is described mathematically for example by cumulative distribution functions, densities, Laplace transforms, moment generating functions, hazard rates or other functionals. Therefore we may apply the *partial order* concept to each of the equivalent descriptions of random variables. (For definitions see Appendix)

Comparisons of functionals of two random variables result sometimes in *partial orders* of random variables. However, frequently we define some relationship between functionals of random variables which does not satisfy the anti-symmetry property of a *partial order*. In this case we obtain a *semi-partial order*, which still may be useful. We call all orderings of random variables *stochastic orderings*. This terminology is also freely applied to *distributions* instead of random variables. By a *distribution* we mean the probability measure generated by a given random variable, which can be described by the corresponding cumulative distribution function or equivalently by the corresponding Laplace transform or by other functionals of a random variable.

From the technical point of view, we can group all random variables with the same distribution into one equivalence class, and then compare their distributions; in fact, we usually compare distributions, not particular versions of random variables corresponding to those distributions. In such situations we do not have to assume that the compared random variables are defined on the same probability space. However, the question whether we can construct versions of the compared random variables that satisfy some additional properties on the same probability space is of interest. Such constructions utilize random variables which are dependent in the probabilistic sense.

One of the most useful stochastic orderings is one that implies almost sure comparison of the corresponding versions of the underlying random variables. In this and in the next chapter we present some constructions of almost surely comparable versions of multivariate random variables, which are random elements of R^n, R^∞, or of other multidimensional spaces. Such constructions can be useful in simulation. They also allow us to prove that the appropriately constructed random variables are stochastically ordered in the strong sense.

1.1 Construction of iid random variables

The fundamental concept of probability theory is the concept of probabilistic or statistical independence. It is our belief that the statistical independence corresponds in a way to traditional or intuitive meanings that two events are "independent". The rule of multiplication of probabilities of independent events is an attempt to formalize this "independence" . One is naturally inclined to believe that events which seem to be unrelated are "independent". Thus considering two experiments taking place in two samples far removed from each other one will readily invoke the rule of multiplication of probabilities (how can they be dependent if one sample is in College Station, Texas, say, the other in New York?). Unfortunately what is involved here is not a strict logical implication. It is our belief that the definition is applicable to a particular situation. There is "independence" in an intuitive sense, and there is the mathematical rule of multiplication, which is narrow, but well defined. We will concentrate ourselves on the mathematical concept of independence, following an observation made by E. Borel that the binary digits are independent. We start with basic definitions (for a mathematical treatment of independence, see Kac (1959)).

Let $(\Omega, \mathcal{F}, \mathbf{P})$ be a probability space. Events $A_1, \ldots, A_n \in \mathcal{F}, n > 1$, are **independent** if

$$\mathbf{P}(A_1 \cap \ldots \cap A_n) = \mathbf{P}(A_1) \cdots \mathbf{P}(A_n).$$

Let $(\Omega, \mathcal{F}, \mathbf{P})$ be a probability space and $X_1, \ldots, X_n$ random variables on Ω. $X_1, \ldots, X_n$ are **independent** if for all $a_1, \ldots, a_n \in R$

$$\mathbf{P}(X_1 \le a_1, \ldots, X_n \le a_n) = \mathbf{P}(X_1 \le a_1) \ldots \mathbf{P}(X_n \le a_n).$$

Denote the vector of random variables $(X_1, \ldots, X_n)$ by $\mathbf{X}$. Then from the definition $X_1, \ldots, X_n$ are independent if and only if

$$\mathbf{F_X}(\mathbf{t}) = F_{X_1}(t_1) \ldots F_{X_n}(t_n),$$

where $\mathbf{t} = (t_1, \ldots, t_n)$, and $\mathbf{F_X}(\mathbf{t})$ is the joint distribution function of $\mathbf{X}$ (see Appendix).

Note that the concept of independence of random variables does not rely on particular versions of the random variables, but is defined with a use of their distributions only.

The following example indicates that the mathematical independence when used to describe an intuitive independence may lead to an unexpected conclusion.

Example. Suppose two persons are playing the following game. The first person is selecting arbitrary two real numbers A and B, say. After tossing a coin, with probabilities 1/2, he displays one of these numbers, and the second person has to guess whether it is the larger one or the smaller one. It is possible to argue using the mathematical independence that the second person can apply an independent procedure that guarantees the right choice with probability grater than 1/2. Our intuitive understanding of independence tells us that the chance of right guessing should be just 1/2. This independent procedure is simply: select a real number at random using the normal distribution, and think of the result as of the hidden number, that is if the result of the random choice is greater than the displayed number, your guess is the displayed number is the smaller one, etc. Employing the mathematical independence we argue as follows. The tossing of a coin we describe by a probability space $\Omega_1 = \{\{A\}, \{B\}\}$ denoting two elementary events of displaying A or B, and the probability P_1 which gives masses 1/2 to both elementary events. The random choice we describe by $\Omega_2 = R$ and the probability measure P_2 generated by the normal distribution function φ on R. The mathematical independence assumed for the tossing and the random choice means that we have to consider the product space and the product probability, $\Omega_1 \times \Omega_2$, $P_1 \otimes P_2$. Now, if we assume $A < B$ to fix our attention, the probability of the right choice is just $\frac{1}{2}(1 - \varphi(A)) + \frac{1}{2}\varphi(B) = \frac{1}{2}(1 + \varphi(B) - \varphi(A)) > \frac{1}{2}$.

A notion often used in probability theory is a sequence of independent random variables, i.e. a sequence for which the random variables in each finite subset of the sequence are independent. One usually assumes the existence of such a sequence, however the proof of the existence of it is in general rather complicated. It is based on the Daniel-Kolmogorov theorem on consistent family of probabilities. We do not use such generality in proving the existence of such a sequence, instead we shall provide a "construction" of the sequence of independent uniformly distributed random variables, from which we shall construct sequences of independent and dependent random variables with prescribed finite dimensional distributions. Such an approach is more useful for stochastic ordering. We shall need in the sequel the following technical lemmas.

Suppose that $F : (a, \infty) \to R \cup \{\infty\}$ is a generalized distribution function (a can be $-\infty$), i.e. it is right continuous and nondecreasing.

The **generalized inverse** function of F is

$$F^{-1}(z) = \inf\{t : F(t) \geq z\},$$

for $z \geq F(a+)$. If the set under the inf sign is empty we take $F^{-1}(z) = \infty$.

Lemma A. *$F^{-1}(z)$ is a nondecreasing function of z , for $z \geq F(a+)$*

Proof. Let $z_1 \geq z_2 \geq F(a+)$. Then

$$\{t; F(t) \geq z_2\} \subseteq \{t : F(t) \geq z_1\},$$

therefore $F^{-1}(z_2) \geq F^{-1}(z_1)$. □

Lemma B. *If $z \geq F(a+)$, $t > a$, then*
(i) $z \leq F(t)$ if and only if $F^{-1}(z) \leq t$,
(ii) $F(t) < z$ if and only if $t < F^{-1}(z)$.

Proof. We prove (i). If $F(t) \geq z$, then $F^{-1}(z) \leq t$, from definition. Assume that $F^{-1}(z) = \inf\{t' : F(t') \geq z\} \leq$t. Then there exists $t' \leq t$ such that $F(t') \geq z$ (the monotonicity of F) or there exist a sequence $\{t_n\}_{n \geq 1}$, such that $t_n \to t$ from the right hand side, and $F(t_n) \geq z, n = 1, \ldots$. In this case we use the fact that F is right continuous. □

Lemma C. *Suppose $F : (a, \infty) \to R$, $G : (b, \infty) \to R$ and*

$$F(t) \geq G(t), \quad t \geq \max\{a, b\}.$$

Then

$$F^{-1}(z) \leq G^{-1}(z), \quad for\ z \geq \max\{F(a+), G(b+)\}.$$

Proof. This is immediate from

$$\{t : G(t) \geq z\} \subseteq \{t : F(t) \geq z\}.$$

□

Lemma D. *Suppose that U is a random variable with the uniform distribution on [0,1]. For an arbitrary distribution function F on R, $F^{-1}(U)$ is a random variable with distribution function F.*

Proof. Here $F^{-1}(z)$ is defined for $z \geq 0$. From Lemma A., it is a monotone function; hence, $F^{-1}(U)$ is a random variable. From Lemma B. we have

$$\mathbf{P}(F^{-1}(U) \leq t) = \mathbf{P}(U \leq F(t)) = F(t), \ t \in R.$$

□

Lemma E. *If t is a point of strict increase for F (i.e. for each $\epsilon > 0$, $F(t-\epsilon) < F(t) < F(t+\epsilon)$), then*

$$F^{-1}(F(t)) = t$$

Proof. Since $F(t) < F(t')$ for all $t < t'$ then

$$F^{-1}(F(t)) = \inf\{t' : F(t') \geq F(t)\} = t.$$

□

Lemma F. *If F is a continuous function then F^{-1} is strictly increasing on $[F(a+), F(\infty)]$.*

Proof. Let $F(a+) \leq z_1 < z_2 \leq F(\infty)$. From the Darboux property for continuous functions we have that there exists t such that $z_1 < F(t) < z_2$, these inequalities are also valid in a neighborhood of t, so

$$\inf\{t' : F(t') \geq z_1\} < t < \inf\{t' : F(t') \geq z_2\}.$$

□

The basic construction we formulate as a theorem.

Theorem G. *Suppose that $\{F_n\}_{n \geq 1}$ is a sequence of distribution functions on R. There exist a probability space and a sequence of independent random variables $\{X_n\}_{n \geq 1}$ on this space such that X_n has the distribution function F_n, $n \geq 1$.*

Proof. As the probability space we take $\Omega = [0,1], \mathcal{F} = \mathcal{B}_{[0,1]}, \mathbf{P} = \ell$ (see Appendix). It is known that every real number $\omega \in [0,1]$ can be written uniquely in the form

$$\omega = Z_1(\omega)/2 + Z_2(\omega)/2^2 + \ldots, \tag{1}$$

where each $Z_i(\omega)$, $i \geq 1$, is either 0 or 1. This is the familiar binary expansion of ω, and to ensure uniqueness we agree to write terminating expansions in the form in which all digits from a certain point on are 0. We prove first that Z_n are independent random variables with the same distribution. Let $n_1 < \ldots < n_k$ be fixed natural numbers, and $\mathbf{a} = (a_1, \ldots, a_k), \mathbf{b} = (b_1, \ldots, b_n)$ vectors with 0-1 valued coordinates. Define

$$A_{\mathbf{a}} = \{\omega : Z_{n_1}(\omega) = a_1, \ldots, Z_{n_k}(\omega) = a_k\}.$$

Consider the following transformation

$$T_{\mathbf{a},\mathbf{b}}(\omega) = \omega + \sum_{i=1}^{k}(b_i - a_i)/2^{n_i},$$

which is a translation on R. It transforms the set $A_{\mathbf{a}}$ onto the set $A_{\mathbf{b}}$. Since $T_{\mathbf{a},\mathbf{b}}$ is a translation, the measures of $A_{\mathbf{a}}$ and $A_{\mathbf{b}}$ are the same, for all arbitrary $\mathbf{a}$ and $\mathbf{b}$. There are 2^k different vectors $\mathbf{a}$, and the corresponding sets $A_{\mathbf{a}}$ are disjoint for different $\mathbf{a}$'s. However all the sets $A_{\mathbf{a}}$ together give the whole interval [0,1]. Hence the measure of each $A_{\mathbf{a}}$ is $1/2^k$. Especially we have $\mathbf{P}(Z_n = 0) = \mathbf{P}(Z_n = 1) = 1/2$, which proves that $Z_n's$ have the same distribution. Also

$$\mathbf{P}(A_{\mathbf{a}}) = \mathbf{P}(\{\omega : Z_{n_1}(\omega) = a_1, \ldots, Z_{n_k}(\omega) = a_k\}) =$$

$$= \mathbf{P}(Z_{n_1}(\omega) = a_1) \ldots \mathbf{P}(Z_{n_k}(\omega) = a_k) = 1/2^k,$$

which implies independence of $Z_{n_k}'s$, for arbitrary $n_1 < \ldots < n_k$.
Now consider a reenumeration of the sequence $\{Z_n\}_{n \geq 1}$ in such a way that we obtain infinite matrix $[Z_{nk}]_{n,k \geq 1}$. For $\omega \in \Omega, n \geq 1$, let

$$U_n(\omega) = \sum_{k=1}^{\infty} Z_{nk}(\omega)/2^k.$$

The above series is convergent for each $\omega \in [0,1]$, and $n \geq 1$. The resulting limit is a random variable U_n. For different n the random variables U_n are independent, because all Z_{nk} are independent. We show that they have the same (uniform) distribution. Consider

$$S_{ni}(\omega) = \sum_{k=1}^{i} Z_{nk}(\omega)/2^k.$$

If $0 \leq m \leq 2^i - 1$ then $\mathbf{P}(S_{ni} = m/2^i) = \mathbf{P}(Z_{n1} = a_1, \ldots, Z_{ni} = a_i) = 1/2^i$, for some $\mathbf{a}$. Thus for a fixed $\omega \in [0,1]$

$$\mathbf{P}(S_{ni} \leq \omega) = ([2^i\omega] + 1)/2^i,$$

where $[2^i\omega]$ denotes the integer part of $2^i\omega$. Since $\{U_n \leq \omega\} = \bigcap_{i=1}^{\infty}\{S_{ni} \leq \omega\}$, and the sequence of sets $\{S_{ni} \leq \omega\}$, $i \geq 1$ is descending, we have finally

$$\mathbf{P}(U_n \leq \omega) = \lim_{i \to \infty} \mathbf{P}(S_{ni} \leq \omega) = \omega,$$

for $\omega \in [0,1]$ and $n \geq 1$.
Now we define $X_n = F_n^{-1}(U_n)$, and from the Lemma D. we know that X_n has the distribution function F_n, $n \geq 1$. □

1.2 Strong ordering

We begin by recalling the notion of a partial order. Let $\mathcal{X}$ be a nonempty set. A binary relation $\leq$ on this set is called **preorder** if
(i) x≤x, x∈ X (reflexivity),
(ii) x ≤y , y≤z ⇒ x≤z (transitivity),
if in addition a preorder has the property
(iii) x≤y, y≤x ⇒ x =y (antisymmetry),
then it is called a **partial order.**
We shall look at several partial orderings on the set $\mathcal{P}$ of all distribution functions on R.

Let $F, G \in \mathcal{P}$. F is **strongly stochastically smaller** than G, written $F <_{st} G$, if $\bar{F}(t) \leq \bar{G}(t)$, $t \in$R.

It is trivial that $<_{st}$ is a partial order on $\mathcal{P}$. In terms of the corresponding random variables we can characterize $<_{st}$ as follows.

Theorem A. *If $F <_{st} G$ then there exist random variables X, Y on the same probability space Ω, with distribution functions $F_X = F$, $F_Y = G$ such that $X(\omega) \leq Y(\omega), \omega \in \Omega$.*

Proof. As the probability space we take $([0,1], \mathcal{B}_{[0,1]}, \ell)$. Then from Lemma 1.1D, we know that $X(\omega) = F^{-1}(\omega)$, and $Y(\omega) = G^{-1}(\omega)$ are random variables with distributions F, and G, respectively. From Lemma 1.1C, and the assumption that $F(t) \geq G(t)$, we have $X(\omega) \leq Y(\omega), \omega \in [0,1]$. □

We write $X <_{st} Y$ if for the corresponding distribution functions $F_X <_{st} F_Y$. If $X <_{st} Y$ then there exist versions (see Appendix) of X and Y, $\tilde{X}, \tilde{Y}$ say, on the same probability space $\tilde{\Omega}$, for which $\tilde{X}(\omega) \leq \tilde{Y}(\omega)$, $\omega \in \tilde{\Omega}$.
The condition $F(t) \geq G(t)$ can be equivalently rewritten as

$$\int_{-\infty}^{\infty} \delta_t(x) dF(x) \leq \int_{-\infty}^{\infty} \delta_t(x) dG(x), \tag{1}$$

where $\delta_t(x)$ denotes the distribution function of the probability measure concentrated at t.
This may be thought to be an artificial way of expressing a simple condition, but it implies the following useful characterization of $<_{st}$.

Theorem B. *$F <_{st} G$ if and only if*

$$\int_{-\infty}^{\infty} \phi(x) dF(x) \leq \int_{-\infty}^{\infty} \phi(x) dG(x) \tag{2}$$

for all increasing functions ϕ, for which the integrals exist.

Proof. Since $\delta_t(x)$ is a nondecreasing function of x for $t \in R$, it is enough to prove the "only if " part of the theorem. Suppose that ϕ is a "simple" function of the form

$$\phi(t) = \sum_{i=1}^{K} a_i \delta_{b_i}(t),$$

for some sequences $\{a_i\}, \{b_i\}$, $i = 1, \ldots, K$, $K \in$N. Then the inequality (2) follows from the additivity properties of integrals and from (1). If ϕ is arbitrary, consider

$$\begin{aligned}\phi_n(t) &= \sum_{k=0}^{n2^n} \delta_{\phi^{-1}(k/2^n)}(t)/2^n + \\ &+ \sum_{k=1}^{n2^n} (\delta_{\phi^{-1}(k/2^n)}(t) - 1)/2^n, \ n \in N.\end{aligned}$$

From this formula we have that $| \phi_n(t) - \phi(t) | \leq 1/2^n$, for $t \in (\phi^{-1}(-n), \phi^{-1}(n))$, hence $\phi_n(t) \to \phi(t)$, as $n \to \infty$ for $t \in$R. In addition we have $\phi_n(t) \leq \phi_{n+1}(t)$, for $t \geq \phi^{-1}(0)$, and $\phi_n(t) \geq \phi_{n+1}(t)$ for $t < \phi^{-1}(0), n \in$N. Now from the monotone convergence theorem for integrals, the proof is complete. □

The above characterization is taken as a definition for the strong stochastic ordering in more dimensional spaces. Taking $\phi(x) = x$, we have for $F <_{st} G$ that

$$\int_{-\infty}^{\infty} x dF(x) \leq \int_{-\infty}^{\infty} x dG(x).$$

This means in terms of the corresponding random variables X, and Y that $EX \leq EY$, where EX, and EY are the expected values of X, and Y, respectively. We have also for $X <_{st} Y$, $\mathrm{E}X^k \leq EY^k, k = 1, 3, 5, \ldots$, and for $k = 1, 2, 3, \ldots$ provided X, and Y are nonnegative.

We can express $<_{st}$ in terms of other functions. For example, a useful function is the following one. Let $F, G \in \mathcal{P}$. The function

$$\psi_{F,G}(t) = G^{-1}F(t), t \in R$$

is a **relative inverse function** of F and G.

Theorem C. *$F <_{st} G$ if and only if $\psi_{F,G}(t) \geq t$*

Proof. Suppose $F <_{st} G$. Since $F(t) \geq G(t)$, $t \in R$, $G^{-1}F(t) = \inf\{s : G(s) \geq F(t)\} = \inf\{s \geq t : G(s) \geq F(t)\} \geq t$. Conversely, suppose $\inf\{s : G(s) \geq F(t)\} \geq t$, then $G(t) \leq F(t)$ i.e. $F <_{st} G$ □

For the life-type distribution functions, i.e. $F(0+) = G(0+) = 0$, which have densities positive on R_+, we have another characterization.

Theorem D. *$F <_{st} G$ if and only if for the corresponding total failure rate functions*

$$R_F(t) \geq R_G(t), \ t > 0.$$

Proof. Recall that for $t > 0$ (see Appendix)

$$R_F(t) = \int_0^t r_F(x) dx,$$

where $r_F(x) = f(t)/\bar{F}(x)$. From this we have

$$\bar{F}(t) = e^{-R_F(t)}, \ t > 0,$$

which applied to F and G gives the desired inequality. □

The last result we can see in another way if we write the formula

$$\bar{G}(t) = e^{-R_G(t)},\ t > 0,$$

in the form

$$R_G(t) = -\ln(1 - G(t)),\ t > 0,$$

and we note that the inverse function for the standard exponential distribution function $F_{\exp}(t) = 1 - e^{-t}$, is equal to

$$F_{\exp}^{-1}(t) = -\ln(1-t),\ t > 0,$$

hence

$$R_G(t) = F_{\exp}^{-1}G(t) = \psi_{G,F_{\exp}}(t), \tag{3}$$

and

$$R_G^{-1}(t) = G^{-1}F_{\exp}(t).$$

Now the relative inverse function of R_F and R_G can be written as

$$R_G^{-1}(R_F(t)) = G^{-1}F_{\exp}R_F(t) = G^{-1}F_{\exp}F_{\exp}^{-1}F(t) = \psi_{F,G}(t). \tag{4}$$

The formula (4) says that the relative behavior of the total failure rates is the same as that of the corresponding distribution functions.
Now the relation $R_F(t) \geq R_G(t)$ is equivalent to $\psi_{F,G}(t) \geq t$ from Theorem 1.2.C.

The construction of a random variable with an arbitrary distribution from a uniform random variable (Lemma 1.1D) can be put in a more general setting. With use of relative inverse functions, we can theoretically construct random variables with given distributions from, for example, exponentially distributed or other random variables.

Let X, and Y be arbitrary random variables with distribution functions F, and G, respectively. Recall that t is a point of increase for F if for each $\epsilon > 0$, $F(t-\epsilon) < F(t+\epsilon)$. The set of all such points for F we call the **support** of F (in symbols suppF).

Theorem E. *If F is continuous on its support which is an interval of R then*

$$\psi_{F,G}(X) =^d Y,$$

where $=^d$ denotes the equality of the corresponding distribution functions.

Proof. Note first that under our assumptions $F(X) =^d U$, where U is a random variable uniformly distributed on [0,1]. Indeed

$$\mathbf{P}(F(X) \leq x) = \mathbf{P}(X \leq F^{-1}(x)) = F(F^{-1}(x)) = x.$$

Utilizing this fact we have

$$\mathbf{P}(\psi_{F,G}(X) \leq x) = \mathbf{P}(G^{-1}F(X) \leq x) = \mathbf{P}(F(X) \leq G(x)) = G(x),$$

which proves the theorem . □

If G is continuous on its support which is an interval of R then

$$\psi_{F,G}^{-1}(z) = \psi_{G,F}(z),$$

which follows from the following equalities

$$\psi_{F,G}^{-1}(z) = \inf\{t : \psi_{F,G}(t) \geq z\} = \inf\{t : G^{-1}F(t) \geq z\} = \inf\{t : F(t) \geq G(z)\} = F^{-1}G(z).$$

Consequently we have

Theorem F.

(i) *If $X = U$, where U is uniformly distributed on [0,1], then*

$$\psi_{F_U,G}(U) =^d Y,$$

for arbitrary Y. (Here $\psi_{F_U,G}(U) = G^{-1}$, hence this case corresponds to Lemma 1.1.D., $=^d$ denotes equality in distribution).

(ii) *If $Y = M$, where M is exponentially distributed, i.e. $G = F_{\exp}$, then*

$$\psi_{F_{\exp},F}(M) =^d X,$$

for arbitrary $X \geq 0$.
(This implies in view of $\psi_{F_{\exp},F}(t) = \psi_{F,F_{\exp}}^{-1}(t)$, that $R_F^{-1}(M) =^d X$.)

To summarize, we can construct arbitrary nonnegative random variables from exponential random variables, using the inverse function of the corresponding total failure rate function. In general, one can construct arbitrary random variables from other random variables if the corresponding relative inverse functions are known and are smooth enough, as in the case when one of the random variables is uniform or exponential.

PROBLEMS AND REMARKS

A. Using a suitable characterization, show that the following families of distributions (see Appendix for definitions) are strongly stochastically ordered

(1) exponential with λ;

(2) Poisson with λ;

(3) Gamma with λ, and with β;

(4) Normal with m;

(5) Weibull with λ.

B. Find random variables X, Y such that $X <_{st} Y$ and $VarX > VarY$.

1.3 Convex ordering

Let us recall the definition of a convex function. Denote by I an arbitrary interval of R (possible infinite). A function $f : I \to R$ is **convex** on I if

$$f(\alpha x + (1-\alpha)y) \leq \alpha f(x) + (1-\alpha)f(y),$$

for all $x, y \in I, \alpha \in [0,1]$.

A useful characterization of convex functions is the following one. A function f is convex on I if and only if for each $c \in I$ there exists an increasing function g such that

$$f(x) - f(c) = \int_c^x g(t)dt, \qquad x \in \mathrm{I}.$$

(Roberts and Varberg (1965)).

Let $F, G \in \mathcal{P}$. F is smaller with respect to the **increasing convex ordering** than G, written $F <_{icx} G$, if

$$\int_{-\infty}^{\infty} \phi(t)dF(t) \leq \int_{-\infty}^{\infty} \phi(t)dG(t),$$

for all increasing convex functions ϕ, for which the integrals exist.

The relation $<_{icx}$ is a partial order on the subset of $\mathcal{P}$ of distributions with finite means. We write $X <_{icx} Y$ if X, and Y have distribution functions related by $F_X <_{icx} F_Y$.

A useful characterization of $<_{icx}$ is the one using the following special convex functions

$$c_x(t) = \int_{-\infty}^{t} \delta_x(s)ds, \ t, x \in R,$$

(In other notation $c_x(t) = (t-x)_+$, where $y_+ = \max(0, y)$, for $y \in R$).

Theorem A. *$F <_{icx} G$ if and only if*

$$\int_{-\infty}^{\infty} c_x(t)dF(t) \leq \int_{-\infty}^{\infty} c_x(t)dG(t), \tag{1}$$

for all $x \in R$.

Proof. We prove the "only if" part of the theorem. Let us use the above representation for an increasing convex function ϕ for which $\phi(-\infty) = 0$, i.e.

$$\phi(t) = \int_{-\infty}^{t} g(s)ds,$$

for some increasing function g.
Note that from the integration by parts formula (Billingsley (1986), p.240) we have

$$\int_{-\infty}^{\infty} c_x(t)dg(x) = \int_{-\infty}^{t} g(x)dx,$$

hence

$$\int_{-\infty}^{\infty} \phi(t)dF(t) = \int_{-\infty}^{\infty}\int_{-\infty}^{t} g(x)dxdF(t) = \int_{-\infty}^{\infty}\int_{-\infty}^{\infty} c_x(t)dg(x)dF(t) =$$

(by Fubini's theorem)

$$= \int_{-\infty}^{\infty}\int_{-\infty}^{\infty} c_x(t)dF(t)dg(x).$$

Thus, whenever we have (1) , we also have (2) for ϕ increasing convex with $\phi(-\infty) = 0$. The general case follows from a standard truncation argument. □

Condition (1) can be rewritten in the following equivalent forms
(i) $E(X - x)_+ \leq E(Y - x)_+,\ x \in R$, where X, Y have the distribution functions F, G, respectively.
(ii) $E\max(x, X) \leq E\max(x, Y),\ x \in R$,
(iii) $\int_x^\infty \bar{F}(t)dt \leq \int_x^\infty \bar{G}(t)dt,\ x \in R$.

An important special case of this ordering is when F and G have the same mean value, i.e. $\int_{-\infty}^{\infty} xdF(x) = \int_{-\infty}^{\infty} xdG(x) = m$. In this case we write $F <_{cx} G$, because the characterizing inequality holds for all convex functions. Indeed, we have

Theorem B. *$F <_{cx} G$ if and only if*

$$\int_{-\infty}^{\infty} \phi(x)dF(x) \leq \int_{-\infty}^{\infty} \phi(x)dG(x), \tag{2}$$

for all convex ϕ, for which the integrals exist.

Proof. The proof of the "only if" part is immediate. Suppose $F <_{cx} G$. Take an arbitrary convex function ϕ. Note that we can decompose ϕ into two functions, one of which is increasing convex, the other one is decreasing convex. Thus it is enough to show (2) for all ϕ decreasing convex. This will be easier to see in terms of random variables. Suppose that X, and Y are random variables with the distribution functions F, and G , respectively. The condition (2) for all ϕ decreasing convex is equivalent to $-X <_{icx} -Y$ since for each ϕ decreasing convex, $\phi(-x)$ is increasing convex. Random variables can be decomposed into two parts: negative and positive, in particular for $X - x$, where x is a real constant we have

$$X - x = (X - x)_+ - (X - x)_-,$$

where $(X - x)_+ = \max(0, X - x)$, $(X - x)_- = \max(0, -(X - x))$. Hence $(X - x)_+ = X - x + (X - x)_-$, and taking expectations $E(X - x)_+ = m - x + E(X - x)_-,\ x \in R$, but $(X - x)_- = (-(X - x))_+,\ x \in R$, thus $E(X - x)_+ = m - x + E(x - X)_+,\ x \in R$. Using (i) after Theorem A., we see that $X <_{icx} Y$ is equivalent to $-X <_{icx} -Y$ in the case of equal mean values. This completes the proof. □

If $X <_{cx} Y$ then $VarX \leq VarY$, and $EX^k \leq EY^k,\ k = 2, 4, \ldots$. If in addition $X \geq 0$, and $Y \geq 0$ then $EX^k \leq EY^k,\ k \geq 1$.

In the case of nonnegative random variables with the same expectation we can relate $<_{cx}$ ordering to $<_{st}$. Suppose F is a distribution function with the finite mean value

$m_F = \int_0^\infty x dF(x)$, and its support in R_+ (i.e. a distribution function of a nonnegative random variable). The **stationary renewal distribution** of F is

$$\tilde{F}(t) = \frac{1}{m_F}\int_0^t \bar{F}(s)ds, \; t \geq 0.$$

From (iii) after Theorem A, we see that $F <_{cx} G$ if and only if $\tilde{F} <_{st} \tilde{G}$.

An alternative method of comparing distributions with the same mean values is by a stochastic transformation. A **stochastic transformation** is a function $T(x, E)$, $x \in R$, $E \in \mathcal{B}^1$, such that for each fixed $x \in R$, $T(x, \cdot)$ is a probability measure on $(R, \mathcal{B}^1)$, and for each fixed $E \in \mathcal{B}^1$, $T(\cdot, E)$ is a function such that $\{x : T(x, E) \leq a\} \in \mathcal{B}^1$, $a \in [0, 1]$ (i.e. $T(\cdot, E)$ is a measurable function or in another language $T(\cdot, E)$ is a random variable on $(R, \mathcal{B}^1)$). We call such transformations **Markov kernels**. T is mean value preserving if the mean value of the probability measure $T(x, \cdot)$ is equal to x, for $x \in R$, i.e. for the corresponding distribution function

$$T(x, y) = T(x, (-\infty, y]), \; y \in R,$$

we have

$$\int_{-\infty}^{\infty} y dT(x, y) = x, \; x \in R.$$

Using T we can obtain another distribution function G from F in the following way

$$G(y) = \int_{-\infty}^{\infty} T(x, (-\infty, y])dF(x).$$

We then write $G = TF$.

The following theorem provides an alternative approach to $<_{cx}$.

Theorem C. (Blackwell (1953)) *Suppose T is a mean value preserving Markov kernel. Then $F <_{cx} G$ if and only if $G = TF$.*

Proof. Suppose $G(y) = \int_{-\infty}^{\infty} T(x, (-\infty, y])dF(x)$ and ϕ is a convex function then

$$\int_{-\infty}^{\infty} \phi(t)dG(t) = \int_{-\infty}^{\infty}\int_{-\infty}^{\infty} \phi(t)T(x, dt)dF(x).$$

From Jensen's inequality for the integral inside and the fact that T is mean value preserving the above expression is greater than

$$\int_{-\infty}^{\infty} \phi(x)dF(x),$$

which gives $F <_{cx} G$.

[1] We sketch the "if" part of the proof for discrete distributions. Suppose that $F <_{cx} G$ and assume in addition that F is a simple distribution, i.e. $F = \sum_{i=1}^n \lambda_i \delta_{x_i}$, for some nonnegative λ_i's such that $\sum_{i=1}^n \lambda_i = 1$ and δ_{x_i}'s denoting deterministic distributions concentrated at x_i's. Consider the following functional

$$\Psi(\mathbf{f}) = \sum_{i=1}^n \lambda_i \hat{f}_i(x_i),$$

[1]This part of the proof can be passed by in the first reading

where $\hat{f}_i$ denotes the smallest upper semi continuous concave function greater than f_i, for arbitrary continuous and bounded functions $\mathbf{f} = (f_1, \ldots, f_n)$. This is a sublinear functional. Consider another functional

$$\Psi_0(\dot{\mathbf{f}}) = \int_{-\infty}^{\infty} f(x)dG(x),$$

where $\dot{\mathbf{f}} = (f, \ldots, f)$. Since $F <_{cx} G$, i.e.

$$\int_{-\infty}^{\infty} f(x)dF(x) \leq \int_{-\infty}^{\infty} f(x)dG(x)$$

we have also

$$\int_{-\infty}^{\infty} f(x)dG(x) \leq \int_{-\infty}^{\infty} \hat{f}(x)dF(x)$$

which for our F means that $\Psi_0 \leq \Psi$ on $\{(f, \ldots, f), f \text{ continuous and bounded}\}$. From the Hahn-Banach theorem, Ψ_0 can be extended to a functional $\tilde{\Psi}_0$ for all $\mathbf{f}$, which, accordingly to Riesz's theorem, is for f_i's tending to zero at infinities, of the form

$$\tilde{\Psi}_0(\mathbf{f}) = \sum_{i=1}^{n} \lambda_i \int_{-\infty}^{\infty} f_i(t)dG_i(t),$$

for some probability distributions G_i's. We define the corresponding mean value preserving Markov kernel as $T(x_i, (-\infty, y]) = G_i(y)$ for the given x_i's and zero elsewhere and we have

$$\begin{aligned} \int_{-\infty}^{\infty} T(x, (-\infty, y])dF(x) &= \sum_{i=1}^{n} \lambda_i T(x_i, (-\infty, y]) \\ &= \sum_{i=1}^{n} \lambda_i G_i(y) = G(y) \end{aligned}$$

which completes the proof for a simple F.
The proof for general F's is beyond the scope of this text (see Alfsen (1971)). □

An interesting and new insight to the case of $<_{cx}$ ordering is given by the notion of a **fusion** of a probability measure. Let us start with a simple example from Elton and Hill (1992).

Suppose P is the discrete distribution with masses 1/6, 1/3, 1/2 at α, β and γ, respectively. Many other distributions may be obtained irreversibly from P by fusing parts of P. For example, if all of the components of P are mixed together, the resulting distribution is a single atom of mass 1 at $\alpha/6 + \beta/3 + \gamma/2$; or if only half of the α atom is fused with half of β atom, the resulting distribution is with masses 1/12, 1/6, 1/2 and 1/4 at α, β, γ and $(\alpha + 2\beta)/3$, respectively. Each fusion may itself be further fused, resulting in still another distribution.

Fusions of a distribution are studied in Elton and Hill (1992) for general spaces. For the needs of the present paragraph we define fusions for probability measures on R.

Suppose P is a probability on $(R, \mathcal{B}^1)$. Recall that for $A \in \mathcal{B}^1$, such that $P(A) > 0$ we define the conditional probability measure by

$$P(B \mid A) = P(A \cap B)/P(A), \ \ B \in \mathcal{B}^1.$$

Of course $P(\cdot \mid A)$ is then a probability measure on $(A, A \cap \mathcal{B}^1)$. Denote the corresponding distribution function by $F(t \mid A)$. The mean value of the distribution $P(\cdot \mid A)$ on A we call the P-barycenter of A, written $b(A, P)$, i.e.

$$b(A,P) = \int_{-\infty}^{\infty} x dF(x \mid A).$$

A probability measure P on $(R, \mathcal{B}^1)$ is an **elementary fusion** of Q if there exists $A \in \mathcal{B}^1$ with $Q(A) > 0, b(A,Q) < \infty$, and $p \in [0,1]$ such that

$$P = Q \mid_{A^c} + (1-p)Q \mid_A + pQ(A)\delta_{b(A,Q)},$$

where δ_x is the atom measure corresponding to the distribution function $\delta_x(t), x \in R$. An elementary fusion takes a fraction p of the mass of a set A and collapses it into the barycenter of A, decreasing proportionally the measure of A elsewhere.

Example. Consider $(R, \mathcal{B}^1, P_{\exp})$. δ_1 is an elementary fusion of $P_{\exp}$; or take $A := (0,5), p = 1/3$. Elementary fusion is the mixed (discrete-continuous) distribution with single atom mass $(1-e^{-5})/3$ at $(1-6e^{-5})/(1-e^{-5})$, with density $2e^{-x}/3$ on (0,5) and density e^{-x} on $(5, \infty)$.

P is a **simple fusion** of Q if there exist a finite sequence $\{P_j\}_{j=0}^n$ of consecutive elementary fusions such that $P_0 = Q$, $P_n = P$, and P_{j+1} is an elementary fusion of P_j, $j = 0, \ldots, n-1$.
P is a **fusion** of Q, written $P \in \mathcal{F}(Q)$, if there exist a sequence $\{P_n\}_{n=1}^{\infty}$ of simple fusions of Q such that for the corresponding distribution functions

$$F_n(t) \to F(t), \; n \to \infty,$$

i.e. $P_n \to_w P$, as $n \to \infty$ (weak convergence).

If Q is purely discrete with exactly two atoms, then $\mathcal{F}(Q)$ consists of all Borel probability measures which have the same mean value and which have support contained in the closed line segment connecting the two atoms.

Consider two probability measures P, and Q on $(R, \mathcal{B}^1)$ with the corresponding distribution functions F, and G, respectively, and the same finite mean. The following result relates the concepts of fusion and convex ordering.

Theorem D. (Elton and Hill (1992)) *$F <_{cx} G$ if and only if $P \in \mathcal{F}(Q)$.*

Proof. Suppose that P is an elementary fusion of Q , i.e.

$$P = Q \mid_{A^c} + (1-p)Q \mid_A + pQ(A)\delta_{b(A,Q)},$$

for some $A \in \mathcal{B}^1$ with $Q(A) > 0, b(A,Q) < \infty$, and $p \in [0,1]$. Let ϕ be an arbitrary convex function. We have

$$\int_{-\infty}^{\infty} \phi(x)dF(x) = \int_{A^c} \phi(x)dG(x) + (1-p)\int_A \phi(x)dG(x) + pQ(A)\phi(b(A,Q)) \qquad (3)$$

Now

$$b(A,Q) = \int_{-\infty}^{\infty} x dG(x \mid A) = \int_A x dG(x)/Q(A),$$

so from Jensen's inequality

$$\phi(b(A,Q)) \leq \int_A \phi(x) dG(x)/Q(A).$$

Applying (3) and Theorem B. we have $F <_{cx} G$. For arbitrary fusions we apply a standard limiting argument.

[2] For the proof of the "if" part of the theorem assume that $F <_{cx} G$. From Theorem C., there exist a Markov kernel T for which

$$\int_{-\infty}^{\infty} y dT(x,y) = x, \ x \in R. \tag{4}$$

and

$$G(y) = \int_{-\infty}^{\infty} T(x, (-\infty, y]) dF(x).$$

Using the standard construction from Chapter 2 to the kernel T, we can construct two random variables X, Y on some probability space with a probability measure **Pr** such that

$$\mathbf{Pr}(Y \in B \mid X = x) = T(x, B),$$

and

$$\mathbf{Pr}(X \leq x) = F(x),$$

for Borel sets B, $x \in R$. (For conditioning see Chapter 2, Section 3). From (4) it is clear that $E(Y \mid X = x) = x$, therefore $E(Y \mid X) = X$ a.s. (This means that (X, Y) forms a martingale, see Section 2.5.). We show that $P(\cdot) = \mathbf{Pr}(X \in \cdot)$ is a fusion of $Q(\cdot) = \mathbf{Pr}(Q \in \cdot)$. Let $\{\pi_j, j \geq 1\}$ be a partition of R into intervals of the length $1/n$, for a fixed n. Define $A_i = X^{-1}(\pi_i)$, $B_i = Y^{-1}(\pi_i), i \geq 1$, and take N big enough to have

$$P(\bigcup_{i=1}^{N} \pi_i) Q(\bigcup_{i=1}^{N} \pi_i) > 1 - \frac{1}{n}.$$

Let $t_{ij} = \mathbf{Pr}(A_j \cap B_i)/\mathbf{Pr}(B_i)$ for $\mathbf{Pr}(B_i) > 0$ and zero otherwise. Note that $t_i = \sum_{j=1}^{N} t_{ij} \leq 1$. Consider the following measure

$$\hat{P}_n = \sum_{i=1}^{\infty} (1 - t_i) Q \mid_{\pi_i} + \sum_{j=1}^{N} \delta_{b_j} [\sum_{i=1}^{\infty} t_{ij} Q(\pi_i)]$$

where

$$b_j = \frac{\sum_{i=1}^{\infty} t_{ij} Q(\pi_i) b(\pi_i, Q)}{\sum_{i=1}^{\infty} t_{ij} Q(\pi_i)}.$$

In words, we obtain $\hat{P}_n$ in N steps, by taking in the first step the proportion t_{i1} of the measure Q from each partition set π_i, and putting this mass to one point b_1 which is a proportional mixture of the local barycenters $b(\pi_i, Q)$. After N steps $\hat{P}_n$ as a transformation of Q has N additional atoms.

---

[2] This part of the proof can be omitted in the first reading

Let

$$\tilde{b}_j = \int_{A_j} X d\mathbf{Pr}/\mathbf{Pr}(A_j).$$

Because $E(Y \mid X) = X$ a.s. we have

$$\tilde{b}_j = \int_{A_j} Y d\mathbf{Pr}/\mathbf{Pr}(A_j).$$

Note that $\tilde{b}_j \in \bar{\pi}_j$ since on A_j values of X are contained in π_j. We show now that the distance between b_j and $\bar{\pi}_j$, say $dis(b_j, \bar{\pi}_j)$, is small by checking the distance between b_j and $\tilde{b}_j$,

$$\begin{aligned}
|\tilde{b}_j - b_j| &= |\sum_{i=1}^{\infty}(\int_{A_j\cap B_i} Y d\mathbf{Pr}/\mathbf{Pr}(A_j)) - b_j| \\
&= |\sum_{i=1}^{\infty}\int_{A_j\cap B_i} Y d\mathbf{Pr} - \sum_{i=1} \mathbf{Pr}(A_j \cap B_i) b(\pi_i, Q)/\mathbf{Pr}(A_j)| \\
&\le \sum_{i=1}^{\infty}(\mathbf{Pr}(A_j \cap B_i)/\mathbf{Pr}(A_j)) \mid \int_{A_j\cap B_i} Y d\mathbf{Pr}/\mathbf{Pr}(A_j \cap B_i) - b(\pi_i, Q) \mid < 1/n.
\end{aligned}$$

The last inequality above follows from the fact that the summands inside $|\cdot|$ both belong to $\bar{\pi}_j$ (with diameter smaller than $1/n$), and the rest is smaller than one as a conditional probability.
To complete the proof we show that P is a weak limit of $\hat{P}_n$'s as n tends to infinity. The result will follow then from the fact that for each n, $\hat{P}_n$ is a fusion of Q (as a matrix fusion) (Elton and Hill (1992), Prop. 3.26). We check that

$$\lim_{n\to\infty} \int_R f d\hat{P}_n = P,$$

for all bounded and continuous functions f. We have

$$\int_R f dP = \int_\Omega f(X) d\mathbf{Pr} = \sum_{j=1}^{N} \int_{A_j} f(X) d\mathbf{Pr} + \int_{\bigcup_{j=N+1}^{\infty} A_j} f(X) d\mathbf{Pr},$$

$$\int_R f d\hat{P}_n = \sum_{j=1}^{N} f(b_j)\mathbf{Pr}(A_j) + \sum_{i=1}^{\infty}(1 - t_i) \int_{\pi_i} f dQ.$$

Now

$$\int_{\bigcup_{j=N+1}^{\infty} A_j} f(X) d\mathbf{Pr} \le \sup f \mathbf{Pr}(\bigcup_{j=N+1}^{\infty} A_j) \le (1/n) \sup f,$$

$$\begin{aligned}
\sum_{i=1}^{\infty}(1 - t_i) \int_{\pi_i} f dQ &\le \sup f \sum_{i=1}^{\infty}(1 - t_i) Q(\pi_i) \\
&= (1 - \sum_{j=1}^{N}\sum_{i=1}^{\infty} t_{ij} Q(\pi_i)) \sup f \\
&= (1 - \sum_{j=1}^{N} \mathbf{Pr}(A_j)) \sup f = (1 - \sum_{j=1}^{N} P(\pi_j)) \sup f \le (1/n) \sup f,
\end{aligned}$$

where $\sup f$ denotes the supremum of f on R.
Finally,

$$|\sum_{j=1}^{N}(\int_{A_j} f(X)d\mathbf{Pr} - f(b_j)\mathbf{Pr}(A_j))| =$$

$$|\sum_{j=1}^{N}(\int_{A_j}(f(X) - f(b_j)d\mathbf{Pr})| \leq$$

$$\sum_{j=1}^{N} sup_{A_j}(| f(X) - f(b_j) |)\mathbf{Pr}(A_j),$$

which is arbitrary small because f is uniformly continuous and $dis(b_j, \pi_j) < 1/n$. □

A very useful criterion for $<_{icx}$ is the Karlin -Novikoff (1966) cut-criterion.

Theorem E. *Suppose that for two distribution functions F, G with finite first moments m_F, m_G, respectively, we have $m_F \leq m_G$ and*

$$F(x) \leq G(x), \text{ for } x \leq \xi,$$

$$F(x) \geq G(x), \text{ for } x > \xi,$$

for some $\xi \in R$ then

$$F <_{icx} G.$$

Proof. Let U be a uniformly distributed random variable on [0,1]. Take $X = F^{-1}(U)$, $Y = G^{-1}(U)$. From Lemma 1.1.D., we know that X has distribution function F, and Y has distribution function G. From our assumptions we have

$$\max(r, X) \leq \max(r, Y), \ r > \xi,$$

and

$$\min(r, X) \geq \min(r, Y), \ r \leq \xi,$$

so

$$E\max(r, X) \leq E\max(r, Y), \ r > \xi.$$

In order to get $<_{icx}$ it is enough to show that

$$E\max(r, X) \leq E\max(r, Y), \ r \in R,$$

To obtain this inequality, note that for all $x, y \in R$ we have the following identities

$$\max(x, y) = \max(0, y - x) + x$$

$$\min(x, y) = \min(0, y - x) + x$$

$$\max(0, x) = -\min(0, -x).$$

Hence

$$\max(r, X) = \max(0, X - r) + r =$$

$$-\min(0, r - X) + r = -\min(X, r) + r + X.$$

Since $E\min(r,X) \geq E\min(r,Y),\ r \leq \xi$ and $m_F \leq m_G$, we have

$$E\max(r,X) \leq E\max(r,Y),\ r \leq \xi.$$

This completes the proof. □

A companion ordering of convex ordering is a concave ordering. Let $F, G \in \mathcal{P}$. F is smaller with respect to the **increasing concave ordering** than G, written $F <_{icv} G$ if

$$\int_{-\infty}^{\infty} \phi(t)dF(t) \leq \int_{-\infty}^{\infty} \phi(t)dG(t),$$

for all increasing concave functions ϕ, for which the integrals exist.

This is a partial ordering on the set of distribution functions with finite mean. Since for each increasing convex function ϕ on R, the function $-\phi(-x)$ is increasing concave, we have $X <_{icx} Y$ if and only if $-Y <_{icv} -X$. Hence properties of one ordering can be translated to the corresponding properties of the other. In the case $EX = EY$, $<_{icx}$ is a reversal of $<_{icv}$ and vice versa, i.e. $X <_{icx} Y$ if and only if $Y <_{icv} X$.

Suppose F, G have finite mean values. The following conditions are equivalent
(i) $F <_{icv} G$,
(ii) $E\min(x,X) \leq E\min(x,Y),\ x \in R$, where X, Y have distribution functions F, G, respectively,
(iii) $\int_{-\infty}^{x} F(t)dt \geq \int_{-\infty}^{x} G(t)$.
Another ordering for life distribution functions is the following one. Suppose that for $F, G \in \mathcal{P}, F(0) = G(0)$. F is smaller than G with respect to **Laplace-Stieltjes transform ordering**, written $F <_L G$, if

$$\int_0^{\infty} e^{-\lambda x}dF(x) \geq \int_0^{\infty} e^{-\lambda x}dG(x),$$

for all $\lambda > 0$.

The $<_L$ relation is a partial ordering on the set of distribution functions with support in R_+, because Laplace-Stieltjes transforms uniquely determine distribution functions.

A characterization of $<_L$ can be given by means of a special class of functions. The following theorem is from Reuter and Riedrich (1981).

Theorem F. *$F <_L G$ if and only if*

$$\int_0^{\infty} \phi(t)dF(t) \leq \int_0^{\infty} \phi(t)dG(t),$$

for all real functions ϕ on R_+, for which

$$(-1)^{n+1}\phi^{(n)}(t) \geq 0, n = 1, 2, \ldots, t > 0$$

($\phi^{(n)}$ denotes n-th derivative of ϕ)

From this theorem we see that
(i) $F <_{icv} G$ implies $F <_L G$,
(ii) $F <_L G$ implies $m_F \leq m_G$ provided the mean values exist.

PROBLEMS AND REMARKS

A. Let X, Y be random variables with the same expectation. $X <_{cx} Y$ iff $E \mid X - a \mid \leq E \mid Y - a \mid$, for all $a \in R$.

B. Let X, Y be independent random variables with the same expectation. $X <_{cx} Y$ iff $E(\psi(X,Y)) \leq E(\psi(Y,X))$ for all ψ such that $\psi(x,y) - \psi(y,x)$ is convex in x for all $y \in R$.

C. Let X, Y be nonnegative with equal expectations. $X <_{cx} Y$ iff $L_X(u) \geq L_Y(u)$, $u \in [0,1]$, where $L_X(u) = \int_0^u F^{-1}(x)dx / \int_0^1 F^{-1}(x)dx$, for F the distribution function of X, similarly for Y. (L_X is called the Lorenz function).

D. If X, Y are independent with finite expectations then $XEY <_{cx} XY$.

E. If (X_n) is an i.i.d. sequence then $\frac{1}{n}\sum_{i=1}^{n} X_i <_{cx} \frac{1}{n-1}\sum_{i=1}^{n-1} X_i$

F. If (X_n) is an i.i.d. sequence then $\frac{1}{n}\sum_{i=1}^{n} a_i \sum_{i=1}^{n} X_i <_{cx} \sum_{i=1}^{n} a_i X_i$, for real a_i's.

G. Suppose $\{F_\alpha\}_{\alpha>0}$ is a family of distribution functions with the common mean m, and α is a scale parameter i.e. $F_\alpha(t-m) = F_1((t-m)/\alpha)$, $\alpha > 0$, $t \in R$, then $\alpha_1 \leq \alpha_2$ implies $F_{\alpha_1} <_{icx} F_{\alpha_2}$.

H. Prove that for normal distribution functions F_1, F_2 with means m_i, and variances σ_i^2, $i = 1,2$, if $m_1 \leq m_2$ and $\sigma_1^2 \leq \sigma_2^2$ then $F_1 <_{icx} F_2$.

I. Prove that for two Weibull distribution functions F_1, F_2 with $m_{F_1} = m_{F_2}$, $\alpha_1 \geq \alpha_2$, we have $F_1 <_{icx} F_2$ (hint: consider failure rates) .

J. Prove that for gamma distribution functions F_1, F_2, if $\beta_1 \geq \beta_2$ and $m_{F_1} \leq m_{F_2}$ then $F_1 <_{icx} F_2$.

1.4 Conditional orderings

We have proved in the previous section that the integrals with respect to probability measures for some classes of functions may serve as functionals suitable for defining orderings of distributions. If we think of these integrals as utility measures, a general scheme for comparing two arbitrary distributions, P, and Q say, is to compare $u(P) \leq u(Q)$ for a class of utility functions $u \in \mathcal{U}$. The class $\mathcal{U}$ can be taken rather arbitrarily, not necessarily containing functionals which are of integral form. For example $u(P)$ can be the median of the corresponding distribution function or a class of quantiles. In general, utility functions need not be linear. We put more restricted requirements for such orderings if we introduce in addition a conditioning procedure. Suppose we have a class of sets $\mathcal{C}$, such that $P(A) > 0$, $Q(A) > 0$ for $A \in \mathcal{C}$. Now let us define $P <_{\mathcal{U},\mathcal{C}} Q$ if $u(P(. \mid A)) \leq u(Q(. \mid A))$ for $u \in \mathcal{U}$, and $A \in \mathcal{C}$ where $P(. \mid A)$ denotes the conditional probability on the set A. For example taking for $\mathcal{U}^1$ the class of functionals of the form $u(P) = \int_{-\infty}^{\infty} f(s)dF(s)$, where F is the corresponding distribution function, f is non-decreasing and $\mathcal{C}^1 = \{\Omega\}$, we obtain $P <_{\mathcal{U}^1,\mathcal{C}^1} Q$ is equivalent to $F <_{st} G$ for the corresponding distribution functions (if f is in addition convex or concave, we get $<_{icx}$, and $<_{icv}$, respectively).

In this section we apply conditioning procedures to $<_{st}$, i.e. to the above described class $\mathcal{U}^1$, with different classes $\mathcal{C}$. Of course we obtain sronger orderings than $<_{st}$ in this way.

First, we introduce so called residual life distribution functions. Suppose F is a life distribution function i.e. $F(0) = 0$. If $t > 0$ is fixed, the distribution function given by

$$F_t(x) = \frac{F(t+x) - F(t)}{1 - F(t)} \delta_0(x), \ x \in R$$

if $1 - F(t) > 0$, and 1 if $1 - F(t) = 0$, is a **residual life distribution function.**. The name "residual life" can be explained by the equality

$$F_t(x) = \mathbf{P}(X \le x + t \mid X > t), x \ge 0$$

where X is a random variable with the distribution function F.

Consider two probability distributions **P**, and **Q** with supports R_+ and corresponding distribution functions F, and G. For technical convenience, we assume that F, and G have continuous failure rates r_F, r_G . We obtain a useful ordering by using $\tilde{\mathcal{C}} = \{(t,\infty), \ t \in R_+\}$ and $\mathcal{U}^1$.

Theorem A. *The following conditions are equivalent*
i) $\mathbf{P} <_{\mathcal{U}^1,\tilde{\mathcal{C}}} \mathbf{Q}$
ii) $F_t <_{st} G_t, \qquad t \ge 0$
iii) $r_F(t) \ge r_G(t), \qquad t \ge 0.$

Proof. *i)* $\Leftrightarrow$ *ii)* From the definition, $\mathbf{P} <_{\mathcal{U}^1,\tilde{\mathcal{C}}} \mathbf{Q}$ if the distribution functions of $\mathbf{P}(\cdot \mid A)$ and $\mathbf{Q}(\cdot \mid A)$ are strongly stochastically ordered for $A \in \tilde{\mathcal{C}}$. For fixed $t \ge 0$

$$\mathbf{P}((-\infty, y] \mid (t,\infty)) = \mathbf{P}((-\infty, y] \cap (t,\infty))/\mathbf{P}((t,\infty)), \ y \in R$$

and if $y > t$,

$$\mathbf{P}((-\infty, y] \mid (t,\infty)) = \mathbf{P}((t, y])/\mathbf{P}((t,\infty)) = \frac{F(y) - F(t)}{1 - F(t)}.$$

Hence

$$\mathbf{P}((-\infty, y] \mid (t,\infty)) = F_t(y - t).$$

Now

$$\mathbf{P}((-\infty, y] \mid (t,\infty)) \ge \mathbf{Q}((-\infty, y] \mid (t,\infty))$$

is equivalent to

$$F_t(x) \ge G_t(x), \ x \in R, \tag{1}$$

i.e.

$$F_t <_{st} G_t.$$

ii)$\Leftrightarrow$iii) The condition (1) can be rewritten as

$$\frac{\bar{F}(t+x)}{\bar{F}(t)} \le \frac{\bar{G}(t+x)}{\bar{G}(t)},$$

which expresses that $\bar{F}(t)/\bar{G}(t)$ is nonincreasing. Taking first derivative and using the continuity of f and g, implies that this is equivalent to

$$-f(t)\bar{G}(t)+g(t)\bar{F}(t)\leq 0,\ t\geq 0,$$

i.e.

$$r_F(t)\geq r_G(t),\ t\geq 0.$$

□

Modifications of the above theorem have been obtained without our continuity assumptions and allowing other supports by standard approximation arguments based on the fact that the class of continuous functions with compact support is dense in the class of integrable functions. The above ordering is called in the literature **the failure rate ordering** (in symbols $<_h$), because of iii), or **the residual life ordering** because of ii).

Utilizing a richer class $C^\sigma=\{(a,b)\cup(c,d): a<b<c<d,\ a,b,c,d\in R\}$ we obtain an ordering $<_{\mathcal{U}^1,C^\sigma}$.

Theorem B. *Suppose* **P** *and* **Q** *have distribution functions* F,G *with positive densities* f *and* g, *respectively, with respect to Lebesgue measure on* R. *Then* $\mathbf{P}<_{\mathcal{U}^1,C^\sigma}\mathbf{Q}$ *if and only if* $g(x)/f(x)$ *is nondecreasing in* $x\in R$.

Proof. If $\mathbf{P}<_{\mathcal{U}^1,C^\sigma}\mathbf{Q}$ then

$$\mathbf{P}((y,\infty)\mid A_\epsilon)\leq\mathbf{Q}((y,\infty)\mid A_\epsilon),$$

where $A_\epsilon=(s-\epsilon,s)\cup(t,t+\epsilon)$ for $s<t$, $\epsilon>0$ and $y\in$R. From this we have

$$\frac{F(t+\epsilon)-F(t)}{F(s)-F(s-\epsilon)+F(t+\epsilon)-F(t)}\leq\frac{G(t+\epsilon)-G(t)}{G(s)-G(s-\epsilon)+G(t+\epsilon)-G(t)}$$

which gives

$$\frac{f(t)}{f(s)}\leq\frac{g(t)}{g(s)}$$

for $s<t$, which completes the proof in one direction.

Suppose now that $g(x)/f(x)$ is nondecreasing. Let $A\in C^\sigma$. Consider

$$f_A(t)=f(t)/P(A),$$

$$g_A(t)=g(t)/Q(A),t\in A,$$

which are densities of the corresponding conditional distributions $\mathbf{P}(.\mid A),\mathbf{Q}(.\mid A)$. Of course $g_A(t)/f_A(t)$ is nondecreasing on A. Because f_A,g_A are both probability densities on A , there exist $c\in A$, such that $g_A(c)/f_A(c)=1$, so

$$g_A(s)/f_A(s)\leq g_A(c)/f_A(c)=1\leq g_A(t)/f_A(t),$$

for all $a<s\leq c\leq t\leq$b. Now for $t\geq c$

$$\bar{F}(t\mid A)=\int_t^\infty f_A(x)dx\leq$$

$$\leq \int_t^\infty f_A(x)g_A(x)/f_A(x)dx = \bar{G}(t \mid A),$$

and for $s \leq c$

$$F(s \mid A) = \int_{-\infty}^s f_A(x)dx \geq$$

$$\geq \int_{-\infty}^s f_A(x)g_A(x)/f_A(x)dx = G(s \mid A),$$

which imply together

$$\bar{F}(x \mid A) \leq \bar{G}(x \mid A), \; x \in R,$$

i.e.

$$\mathbf{P}(. \mid A) <_{st} \mathbf{Q}(. \mid A).$$

□

The ordering $\mathbf{P} <_{\mathcal{U}^1, \mathcal{C}^\sigma} \mathbf{Q}$ is called **the monotone likehood ratio ordering** (in symbols $<_{mlr}$). The $<_{mlr}$ ordering is stronger than $<_{st}$. Note that the above theorem can be proved under more general assumptions about state space, densities, and supports, see Whitt (1980).

PROBLEMS AND REMARKS

A. Let (X_n) be a sequence of i.i.d. random variables, which are nonnegative and have logconcave densities then $\sum_{i=1}^n X_i <_{mlr} \sum_{i=1}^{n+1} X_i$, $n \geq 1$. [Keilson and Sumita (1982)].

B. If $X_i <_{mlr} Y$, $i = 1, \ldots, n$ then $X_N <_{mlr} Y$, for every independent random variable N with a discrete distribution on $\{1, \ldots, n\}$.

C. Let (X_n) be a sequence of i.i.d. random variables with an absolutely continuous distribution. If $M_k = \max(X_1, \ldots, X_k)$, and $m_k = \min(X_1, \ldots, X_k)$ then $M_k <_{mlr} M_{k+1}$, $m_{k+1} <_{mlr} m_k$, and $m_k <_{mlr} M_k$, $k \geq 1$.

1.5 Relative inverse function orderings

In this section we restrict our attention to functions ϕ which are continuous, nonnegative, and for which $\phi(0) = 0$. Let ϕ be defined on $[0, \infty)$. The **average function** $\hat{\phi}$ of function ϕ is the function defined for all $x > 0$ by

$$\hat{\phi}(x) = \frac{1}{x} \int_0^x \phi(t)dt.$$

The function ϕ is **starshaped** if for each $\alpha \in [0, 1]$, and all x

$$\phi(\alpha x) \leq \alpha \phi(x).$$

The function ϕ is **superadditive** if

$$\phi(x + y) \geq \phi(x) + \phi(y),$$

for all x,y.

A function ϕ is **convex (starshaped, superadditive) on the average** if $\hat{\phi}$ is convex (starshaped, superadditive).

Simple characterizations of the introduced classes we take from Bruckner and Ostrow (1963):
ϕ is starshaped if and only if $\phi(x)/x$ is increasing in x; ϕ is starshaped on the average if and only if $\phi \geq 2\hat{\phi}$. If ϕ is respectively convex, convex on the average, starshaped or superadditive then ϕ is a nondecreasing function.

Let $\{\phi_n\}$ be a sequence of convex, starshaped, or superadditive functions converging pointwise to a limit function ϕ. Then ϕ is respectively convex, starshaped, or superadditive.

The following relationships hold among the six classes.

Theorem A. (Bruckner and Ostrow (1963)) *Let ϕ be nonnegative, continuous, such that $\phi(0) = 0$. Consider the following conditions on ϕ:*
i) ϕ is convex
ii) ϕ is convex on the average
iii) ϕ is starshaped
iv) ϕ is superadditive
v) ϕ is starshaped on the average
vi) ϕ is superadditive on the average.
Then i)$\Rightarrow$ii)$\Rightarrow$iii)$\Rightarrow$iv)$\Rightarrow$v)$\Rightarrow$vi).

The above six classes may be used to define stochastic orderings when applied to relative inverse functions. Suppose F, and G are life distribution functions ($F(0) = G(0) = 0$) with supports R_+. Then
i) $F <_c G$(F is convex with respect to G) if $\phi_{F,G}$ is convex on R_+
ii) $F <_* G$ (F is starshaped with respect to G) if $\phi_{F,G}$ is starshaped on R_+
iii) $F <_{su} G$ (F is superadditive with respect to G) if $\Phi_{F,G}$ is superadditive on R_+.

These definitions can be extended to distribution functions whose support is an interval (Dharmadhikari and Joag-Dev (1988)).

From Theorem A. we have $F <_c G \Rightarrow F <_* G \Rightarrow F <_{su} G$. Note that the above orderings are unaffected by rescaling $F(\alpha x), G(\beta x)$, by a positive scale factors α, β. Thus in discussing these orderings we may group into equivalence classes $\{F(\alpha x), \alpha > 0\}, \{G(\beta x), \beta > 0\}$ distributions that differ only by a positive scale factor.

Therefore in order to relate the relative inverse orderings to integral orderings, we have to assume that the corresponding mean values are ordered.

Theorem B. *Suppose F, and G are life distribution functions with support R_+ and finite means m_F, and m_G. If $m_F \leq m_G$ and $F <_* G$ then $F <_{icx} G$.*

Proof. From $F <_* G$ we have

$$\phi_{F,G}(\alpha x) \leq \alpha \phi_{F,G}(x), \ \alpha \in (0,1), \ x > 0,$$

which yields

$$\phi_{F,G}(\alpha x)/\alpha x \leq \phi_{F,G}(x)/x.$$

Since $\phi_{F,G}(x)/x$ is nondecreasing, $\phi_{F,G}(x)$ crosses the line $y = x$ at most once from below, at ξ, say. In this case $\phi_{F,G}(\xi) = \xi$ i.e. $F(\xi) = G(\xi)$, and $F(x) > G(x)$ for $x > \xi$, $F(x) < G(x)$, $x < \xi$. Now from the criterion of Karlin and Novikoff we have $F <_{icx} G$. □

If $m_F \leq m_G$ and $F <_c G$ then $F <_{icx} G$.

For life distribution functions F and G with support R_+, let

$$\mathcal{I}_<(G) = \{F : G < F\},$$

$$\mathcal{D}_<(G) = \{F : F < G\},$$

where $<$ denotes one of the relative inverse orderings. For example $\mathcal{D}_{<_c}(G)$ denotes the class of life distribution functions which are convex with respect to G.

Such classes have interesting closure properties if we take G for example a half-normal, Weibull, lognormal, or gamma distribution.

The most useful classes are obtained by taking $G(t)$=Exp(t)= $(1 - e^{-t})\delta_0(t)$, $t \in$R.

A function $p(x, y)$ defined for $x, y \in R$ is **totally positive of order 2** (TP_2) if $p(x, y) \geq 0$, and

$$det \begin{pmatrix} p(x_1, y_1) & p(x_1, y_2) \\ p(x_2, y_1) & p(x_2, y_2) \end{pmatrix} \geq 0$$

whenever $x_1 \leq x_2, y_1 \leq y_2$.

A function $p(x)$ defined for $x \in R$ is a **Polya frequency function of order 2** (PF_2) if $p(x) \geq 0$ and $p(x - y)$ is TP_2 in $x, y \in R$.

The class $\mathcal{D}_{<_c}(Exp)$ we denote by IFR (increasing failure rate), due to the following characterization.

Theorem C. *Suppose F is a life distribution function with a density f positive on R_+. The following statements are equivalent*
i) $F <_c Exp \quad (F \in \mathcal{D}_{<_c}(Exp))$
ii) $r_F(t)$ *is nondecreasing*
iii) $\bar{F}(t)$ *is* PF_2
iv) $\bar{F}(t)$ *is logconcave.*

Proof. *i)* $\Leftrightarrow$ *ii)*. $F <_c Exp$ means $\phi_{F,Exp}(x) = R_F(x)$ is convex, i.e. r_F is nondecreasing.
ii) $\Leftrightarrow$ *iii)* $\bar{F}(t)$ is PF_2, i.e.

$$\bar{F}(x_1 - y_1)\bar{F}(x_2 - y_2) \geq \bar{F}(x_1 - y_2)\bar{F}(x_2 - y_1)$$

for $x_1 \leq x_2, y_1 \leq y_2$. Put $s = x_1 - y_1, t = x_2 - y_1$, and $\Delta = y_2 - y_1$. We have equivalently

$$\bar{F}(s)/\bar{F}(s - \Delta) \geq \bar{F}(t)/\bar{F}(t - \Delta),$$

for $s \le t, \Delta \ge 0$ which means $\bar{F}(s)/\bar{F}(s-\Delta)$ is a nondecreasing function of s. Taking derivative we come up with

$$-f(s)\bar{F}(s-\Delta)+f(s-\Delta)\bar{F}(s)\le 0,$$

i.e.

$$r_F(s-\Delta)\le r_F(s),\ \Delta\ge 0,\ s\in R.$$

i)⇔iv). We have $\phi_{F,Exp}(x)=R_F(x)=-\ln\bar{F}(x)$ is convex. This is equivalent to $\ln\bar{F}(x)$ is concave, i.e. $\bar{F}$ is logconcave. □

The class $\mathcal{I}_{<_c}(Exp)$ we denote by DFR (decreasing failure rate) because of the following characterization.

Theorem D. *Suppose F is a life distribution function with a density f , which is positive on R_+. The following statements are equivalent:*
i) $Exp <_c F \quad (F\in\mathcal{I}_{<_c}(Exp))$
ii) $r_F(t)$ *is nonincreasing*
iii) $\bar{F}(x+y)$ *is* TP_2 *in* x,y *for* $x+y\ge 0$
iv) $\bar{F}(t)$ *is logconvex*

Proof. The proof is similar to that of the previous theorem and we omit it. □

We introduce a special notation for other classes of the type $\mathcal{D}_<(Exp)$, and $\mathcal{I}_<(Exp)$.
1. $\mathcal{D}_{<_*}(Exp)=IFRA \qquad (F\in IFRA \Leftrightarrow F <_* \text{Exp}),$
(increasing failure rate in the average)
2. $\mathcal{D}_{<_{su}}(Exp)=NBU \qquad (F\in NBU \Leftrightarrow F <_{su} \text{Exp}),$
(new better than used)
3. $\mathcal{I}_{<_*}(Exp)=DFRA \qquad (F\in DFRA \Leftrightarrow \text{Exp} <_* F),$
(decreasing failure rate in the average)
4. $\mathcal{I}_{<_{su}}(Exp)=NWU \qquad (F\in NWU \Leftrightarrow \text{Exp} <_{su} F),$
(new worse than used)
5. $\mathcal{D}_{<_{icx}}(Exp)=HNBUE \qquad (F\in HNBUE \Leftrightarrow F <_{icx} \text{Exp}(1/m_F)),$
(harmonic new better than used in expectation)
6. $\mathcal{I}_{<_{icx}}(Exp)=HNWUE \qquad (F\in HNWUE \Leftrightarrow \text{Exp}(1/m_F) <_{icx} F),$
(harmonic new worse than used in expectation).

It is interesting that some of these classes can be characterized by the corresponding residual life time distribution functions.

Theorem E. *$F\in$IFR (DFR) if and only if $F_t <_{st} F_s$ ($F_s <_{st} F_t$) for all $s \le t$.*

Proof. We prove only the IFR case, the DFR case can be obtained analogously. The condition $F_t <_{st} F_s$ means from the definition

$$\bar{F}(x+t)\ /\bar{F}(t)\le \bar{F}(x+s)\ /\bar{F}(s)\ ,$$

for $x\ge 0, s\le$t. This inequality expresses the fact that for each fixed x , $\bar{F}(x+y)/\ \bar{F}(y)$ is a nondecreasing function of $y\ge 0$. Taking derivative, this statement is equivalent to

$$-f(x+y)\bar{F}(y)+f(y)\bar{F}(x+y)\le 0,$$

which means r_F is nondecreasing . □

Theorem F. *$F \in NBU$ (NWU) if and only if $F_t <_{st} F$ $(F <_{st} F_t)$.*

Proof. From the definition $F \in NBU$ if $\phi_{F,Exp}(t)$ is superadditive. We know that $\Phi_{F,Exp}(t) = R_F(t) = -\log \bar{F}(t)$, so

$$-\log \bar{F}(x+t) \geq -\log \bar{F}(x) - \log \bar{F}(t), \;\; x,t \geq 0,$$

which is equivalent to

$$\bar{F}(x+t)/\bar{F}(t) \leq \bar{F}(x), \;\; x,t \geq 0.$$

The last inequality yields from the definition $F_t <_{st} \mathrm{F}$. □

Classes $\mathcal{I}_<(G)$ and $\mathcal{D}_<(G)$ possess a lot of useful and elegant closure properties. We recall only that IFR class is closed under convolutions and DFR class is closed under mixture. A general theory is developed for classes of distributions determined for example by G which is half-normal, Weibull, lognormal, gamma or uniform (see e.g. Leon and Lynch (1983)). The following theorems are from Barlow and Proschan (1981).

Theorem G. *If F and G are IFR then their convolution*

$$H(t) = \int_0^t F(t-x)dG(x)$$

is IFR.

The proof is based on the Basic Composition Formula of Polya and Szegő (1925). Because many of the structural properties are deducible from this formula we recall its formulation.
Consider a real function $K(x,y)$ such that x,y are from linearly ordered subsets of R. For two increasing vectors $\mathbf{x} = (x_1,\ldots,x_n)$, $x_1 < \ldots < x_n$, and $\mathbf{y} = (y_1,\ldots,y_n)$, $y_1 < \ldots < y_n, n \in N$, we write

$$\det K_{\mathbf{x},\mathbf{y}} = \det \begin{bmatrix} K(x_1,y_1) & \ldots & K(x_1,y_n) \\ \vdots & \vdots & \vdots \\ K(x_n,y_1) & \ldots & K(x_n,y_n) \end{bmatrix}.$$

Theorem H. (Basic Composition Formula) *If the integral*

$$K(x,y) = \int L(x,z)M(z,y)dz$$

is absolutely convergent then

$$\det K_{\mathbf{x},\mathbf{y}} = \int \ldots \int_{z_1<\ldots<z_n} \det L_{\mathbf{x},\mathbf{z}} \det M_{\mathbf{z},\mathbf{y}} \; dz_1 \ldots dz_n$$

for all increasing vectors $\mathbf{x}$ *and* $\mathbf{y}$.

Proof of Theorem G. . Let $K(x,y) = \bar{H}(x-y)$. Take $\mathbf{x} = (x_1,x_2)$, $x_1 < x_2$, $\mathbf{y} = (y_1,y_2)$, $y_1 < y_2$. We show that $\det K_{\mathbf{x},\mathbf{y}} \geq 0$, which is from Theorem C. , iv) equivalent to $\bar{H} \in IFR$. From the Basic Composition Formula we have

$$\det K_{\mathbf{x},\mathbf{y}} = \int\int_{z_1<z_2} \det L_{\mathbf{x},\mathbf{z}} \det M_{\mathbf{z},\mathbf{y}} dz_1 dz_2$$

for $L(x,z) = F(x-z)$ and $M(z,y) = g(z-y)$. Integrating with respect to z_2, the inner integral is equal to

$$\int_{z_2>z_1} (\bar{F}(x_1-z_1)\bar{F}(x_2-z_2) - \bar{F}(x_1-z_2)\bar{F}(x_2-z_1))$$
$$(g(z_1-y_1)g(z_2-y_2) - g(z_1-y_2)g(z_2-y_1))dz_2.$$

Applying the integration by parts formula (Billingsley (1986), p.240) we rewrite this in the form

$$\int_{z_2>z_1} (\bar{F}(x_1-z_1)f(x_2-z_2) - f(x_1-z_2)\bar{F}(x_2-z_1))$$
$$(g(z_1-y_1)\bar{G}(z_2-y_2) - g(z_1-y_2)\bar{G}(z_2-y_1))dz_2.$$

Now it is enough to show that the differences under the integral are nonnegative. The first difference is nonnegative if and only if

$$f(x_2-z_2)/\bar{F}(x_2-z_1) \geq f(x_1-z_2)/\bar{F}(x_1-z_1)$$

which is equivalent to

$$r_F(x_2-z_2)\bar{F}(x_2-z_2)/\bar{F}(x_2-z_1) \geq r_F(x_1-z_2)\bar{F}(x_1-z_2)/\bar{F}(x_1-z_1)$$

The above inequality follows from our assumption: r_F is nondecreasing and hence $\bar{F}$ is PF_2 (Theorem C. iii)). A similar argument holds for the second difference. □

Theorem G. can be reformulated as follows. If $F, G \in \mathcal{D}_{<_c}(Exp)$ then $F*G \in \mathcal{D}_{<_c}(Exp)$, i.e. the class $\mathcal{D}_{<_c}(Exp)$ is closed under convolution.

If $F <_c$Exp and $G <_c$Exp then $F*G <_c$Exp, in other words if X, and Y are two independent random variables such that $X <_c M$ and $Y <_c M$ then $X+Y <_c M$, where M has the standard exponential distribution.

The class DFR is not closed under convolutions, but it has another closure property.

Theorem I. *If $\{F_\alpha\}_{\alpha\in R}$ is a family of DFR distribution functions then the mixture of this family according to a distribution function G :*

$$F(x) = \int_{-\infty}^{\infty} F_\alpha(x)dG(\alpha)$$

is a DFR distribution function.

Proof. It is enough to show that the total hazard function R_F for F is concave. From our assumption each R_{F_α} is concave and R_F is related to R_{F_α} by

$$R_F(t) = -\log \int e^{-R_\alpha(t)} dG(\alpha),$$

so we can think of $R_F(t)$ as a function of the family $\{R_{F_\alpha}(t)\}_{\alpha\in R}$, i.e.

$$R_F(t) = \eta((R_{F_\alpha}(t))_{\alpha\in R})$$

where

$$\eta((u_\alpha)_{\alpha\in R}) = -\log \int e^{-u_\alpha} dG(\alpha).$$

The function η can be proved to be concave from the Hölder inequality, and $R_F(t)$ is concave as a composition of η, which is increasing and concave, and the family $\{R_\alpha(t)\}_{\alpha\in R}$, for which all $R_\alpha(t)$ are concave in t (see Lemma 4.6, p.103 in Barlow and Proschan (1975)) . □

PROBLEMS AND REMARKS

A. Find distribution functions F, G which are both IFR and $\alpha F+(1-\alpha)G$, $\alpha \in (0,1)$ is not IFR.

B. Find distribution functions F, G which are both DFR and $F*G(t)=\int_0^t F(t-y)dG(y)$ is not DFR.

C. If $F <_* G$ then $1-F(x)$ crosses $1-G(\alpha x)$ at most once, and from above, as x increases from 0 to ∞, for each $\alpha > 0$.

D. If X is IFR ($IFRA$) then $X_{j:n}$ is IFR ($IFRA$), where $X_{j:n}$ is the jth order statistic in a sample of n from the distribution of X.

E. If X, Y are nonnegative with $EX^s = EY^s$ for a fixed $s > 0$ then $X <_* Y$ implies $\int_0^\infty \psi(x)x^{s-1}(1-F(x))dx \leq \int_0^\infty \psi(x)x^{s-1}(1-G(x))dx$, for all increasing functions ψ, where F, G are the respective distribution functions of X, Y. From this, if F is $IFRA$ then $\lambda_r^{1/r}$ is decreasing in $r \geq 0$, where $\lambda_r = E(X^r)/\Gamma(r+1)$. [Barlow and Proschan (1981)]

F. If F is NBU (NWU) then $\lambda_{r+s} \leq (\geq)\lambda_r\lambda_s$, for all $r, s \geq 0$.

G. If X is NBU then $EX^s < \infty$ for all $s > 0$.

H. If F, G are NBU then the convolution $F(t) = \int_0^t F(t-x)dG(x)$ is NBU. [Marshall and Proschan (1972)].

I. The NWU class is not preserved under mixtures.

J. If $X <_* Y$ and $EX \leq EY$ then $EY(X-EX) <_{icx} EX(Y-EY)$, and thus $VarX \leq VarY$. [Szekli (1987)].

K. The class of discrete DFR distributions is convex. The class of all extreme points in the class of discrete DFR distributions consists of δ_0 (atom at 0) and distributions for which $r(0) = r(1) < 1$ and $r(k)$ is constant for $k \geq 1$ or $k \geq 2$, where r is the corresponding failure rate. [Langberg et al. (1980)].

L. The class of extended decreasing failure rate distributions (i.e., distributions with support in $[0,\infty]$) is convex and compact. [Langberg et al. (1981)].

M. Suppose X is a positive random variable with a positive failure rate $r(t)$. If $\alpha < \liminf_{t\to\infty} tr(t)$ then $E(X^\alpha) < \infty$, and $E(exp(\alpha X)) < \infty$. If $\alpha > \limsup_{t\to\infty} tr(t)$ then $E(X^\alpha) = \infty$, and $E(exp(\alpha X)) = \infty$. Thus IFR distributions have all finite moments of order $\alpha > 0$, and DFR distributions do not have finite moments of order $\alpha > 0$ if for them $tr(t) \to 0$, $t \to \infty$.

1.6 Dispersive ordering

Let F, G be two arbitrary distribution functions. F and G are **ordered in dispersion**, in symbols $F <^{disp} G$, if

$$F^{-1}(\beta) - F^{-1}(\alpha) \leq G^{-1}(\beta) - G^{-1}(\alpha)$$

for all $0 < \alpha < \beta < 1$.

Useful characterizations of this ordering are based on sign change properties of the corresponding distribution functions. We adopt notation $f_c(x) = f(x - c)$ for a real function f with domain $I \subseteq$R. For a function f the number of sign changes in I is defined by

$$S^-(f) = \sup_{x_1 < \ldots < x_n} S^-(f(x_1), \ldots, f(x_n)),$$

where $S^-(y_1, \ldots, y_n)$ is the number of sign changes of the sequence $y_1, \ldots, y_n$, and supremum is over all $n \in$N.

Theorem A. (Shaked (1982)) *Suppose that F, G have strictly increasing and continuous inverses on (0,1) then*

$$F <^{disp} G$$

if and only if

$$S^-(F_c - G) \leq 1, \; c \in R,$$

with a sign sequence -,+ if it exists.

Proof. Assume that $F <^{disp} G$ and $S^-(F_c - G) \leq 1$ does not hold for some $c_0 \in R$, i.e. after a crossing point x_0 at which $F(x_0 - c_0) = G(x_0)$, we have $F(x - c_0) - G(x) < 0$, for $x_0 < x \leq x_0 + \epsilon, \epsilon > 0$. If we take $\alpha = F(x_0 - c_0), \beta = F(x_0 - c_0 + \epsilon)$, we have a contradiction because

$$G^{-1}(\beta) - G^{-1}(\alpha) < \epsilon = F^{-1}(\beta) - F^{-1}(\alpha).$$

To prove the converse part, take arbitrary $\alpha \in (0, 1)$, and the corresponding x_α such that $G(x_\alpha) = \alpha$. Taking $c_\alpha = G^{-1}(\alpha) - F^{-1}(\alpha)$ we have $F_{c_\alpha}(x_\alpha) = \alpha$, i.e. x_α is a crossing point for F_{c_α} and G. From our assumption we have then $F_{c_\alpha}(x) \geq G(x)$ for $x > x_\alpha$. Taking x_β for $\beta > \alpha$ we have $x_\beta > x_\alpha$ hence $F_{c_\alpha}(x_\beta) \geq G(x_\beta) = \beta$. This gives $F(x_\beta - c_\alpha) \geq \beta$ which after taking inverses, and using definition of c_α leads to $G^{-1}(\beta) - F^{-1}(\beta) \geq G^{-1}(\alpha) - F^{-1}(\alpha)$. □

Theorem B. *If F, G are life distributions with supports R_+ then*

$$F <^{disp} G$$

if and only if

$$F <_{st} G,$$

and

$$S^-(F_c - G) \leq 1, \; c > 0,$$

with the sign sequence -,+ if a crossing occurs.

Proof. See Shaked (1982).

The ordering in dispersion can be also characterized by relative inverse functions under usual smoothness properties.

Theorem C. *If F, and G are absolutely continuous life distribution functions with supports R_+ then $F <^{disp} G$ if and only if $\phi'_{F,G}(t) \geq 1$, for $t \geq 0$.*

Proof. Fix$\alpha < \beta$ and take x_α, x_β such that $F(x_\alpha) = \alpha$, $F(x_\beta) = \beta$. We have $\phi_{F,G}(x_\beta) - \phi_{F,G}(x_\alpha) = G^{-1}(\beta) - G^{-1}(\alpha)$. If $\phi'_{F,G} \geq 1$ then $\phi_{F,G}(x_\beta) - \phi_{F,G}(x_\alpha)/(x_\beta - x_\alpha) \geq 1$, which gives $G^{-1}(\beta) - G^{-1}(\alpha) \geq F^{-1}(\beta) - F^{-1}(\alpha)$. If $F <^{disp} G$ then $G^{-1}(F(x_\beta)) - G^{-1}(F(x_\alpha)) \geq x_\beta - x_\alpha$ for every $x_\beta < x_\alpha$, which is $\phi_{F,G}(x_\beta) - \phi_{F,G}(x_\alpha)/(x_\beta - x_\alpha) \geq 1$. Letting $x_\beta \to x_\alpha$ we have $\phi'_{F,G}(x_\alpha) \geq 1$ for arbitrary x_α. □

In many applications the respective densities have simple closed forms and the following sufficient condition for $F <^{disp} G$ can be used.

Theorem D. *Let F, and G be absolutely continuous distributions with the corresponding densities f, and g. If*

$$S^-(f_c - g) \leq 2, \; c \in R,$$

with the sign sequence -,+,- , in case of equality then

$$F <^{disp} G.$$

Proof. If $S^-(f_c - g) \leq 1$ then F_c and G do not cross each other and $S^-(F_c - G) = 0$. Suppose $S^-(f_c - g) = 2$ with the sign sequence -,+,-. Denote by x_{c_1}, x_{c_2} the cut points of the densities. For $x \leq x_{c_1}, F_c(x) \leq G(x)$, for $x > x_{c_1}, F_c$ can remain below G or if f_c is sufficiently big then at some point $y \in (x_{c_1}, x_{c_2})$, $F_c(y) > G(y)$ and it remains $F_c(x) \geq G(y)$ for $x > y$ because otherwise we would have $F_c(\infty) < G(\infty)$. Thus $S^-(F_c - G) \leq 1$, and the sign sequence is -,+. Now from Theorem A. we have $F <^{disp}$G. □

Example. For gamma densities $f^\gamma(x) = \Gamma(\gamma)x^{\gamma-1}e^{-x}\delta_0(x), \gamma > 0$ it can be checked that $S^-(f_c^\gamma - f^\gamma) \leq 2$, for $c > 0$ and $\gamma_1 < \gamma_2$. From Theorem B. and D. and the fact that $F^{\gamma_1} <_{st} F^{\gamma_2}$ for $\gamma_1 < \gamma_2$ we have $F^{\gamma_1} <^{disp} F^{\gamma_2}$. □

PROBLEMS AND REMARKS

A. For each random variable X, and $a \geq 1$, $X <^{disp} aX$.

B. If X has a logconcave density then $X <^{disp} X + Y$, for any independent random variable Y.

C. If X, Y are nonnegative then $X <^{disp} Y$ implies $X <_{st} Y$.

D. If X, Y have finite expectations then $X <^{disp} Y$ implies $X - EX <_{cx} Y - EY$ (hence $VarX \leq VarY$).

E. If $X <^{disp} Y$ then $X_{j:n} <^{disp} Y_{j:n}$ (order statistics). [Bartoszewicz (1986)].

1.7 Compounding

Partial sums of random variables arise naturally in many applied probability models. In this section we recall some general results dealing with comparisons of random sums, which older versions can be found in Borovkov (1976), Ross (1983), Stoyan (1983) and their generalizations in Jean-Marie and Liu (1992), and we shall treat some aging properties and distances to the exponential distribution of geometric compounds in a more detailed way. Some applications in queueing will be pointed out in the next two sections. Related results about renewal processes will be given also in Section 2.10.

Let M and N be two nonnegative integer valued random variables, $\{X_n\}_{n\geq 1}$, $\{Y_n\}_{n\geq 1}$ be sequences of independent random variables. We assume that M and N are independent of the $\{X_n\}$ and $\{Y_n\}$ sequences. We define random sums by

$$U = \sum_{i=1}^{M} X_i, \quad V = \sum_{i=1}^{N} Y_i,$$

where by convention, $U = 0$ if $M = 0$.
The following theorem which follows immediately from Proposition 3.1 in Jean-Marie and Liu (1992) is not hard to prove.

Theorem A.

(i) *If $M <_{st} N$, $X_i <_{st} Y_i$, $i \geq 1$ and $Y_i \geq 0$ a.s. then $U <_{st} V$.*

(ii) *If $M <_{icx} N$, $X_i <_{icx} Y_i$, $i \geq 1$ and either $X_i \geq 0$ a.s., $X_i <_{icx} X_{i+1}$, $i \geq 1$ or $Y_i \geq 0$ a.s., $Y_i <_{icx} X_{i+1}$, $i \geq 1$ then $U <_{icx} V$.*

Geometric compounds

If $\{X_i\}_{i\geq 1}$ is an i.i.d. sequence of nonnegative random variables and N_0 is geometrically distributed, $\mathbf{Pr}(N_0 = i) = \rho^i(1-\rho)$, $i = 0, 1, 2, \ldots$, and independent of $\{X_i\}$, then $C_0 = \sum_{i=0}^{N_0} X_i$, $(X_0 = 0)$ is a **geometric compound** of $\{X_i\}$. Closely related to C_0 is the random variable $C = \sum_{i=1}^{N} X_i$, where $N = N_0 + 1$, which is also referred to as a geometric compound.

Gertsbakh (1984) discusses a rich variety of applications of geometric compounds in reliability and queues and surveys research in the area. Feller (1971) discusses terminating renewal processes, the time until termination being a geometric compound. Jacobs (1986) investigates a geometric compound in the context of combining random loads and waiting for the stress to exceed a given level. In a GI/GI/1 queue in equilibrium, the waiting time distribution is a geometric compound and has been studied in this context by Kőllerstrőm (1976) and Szekli (1986). We shall discuss these and other applications in the next sections.

If the i.i.d. sequence $\{X_i\}$ of nonnegative random variables has a common marginal distribution $F(x)$ then C_0 has a distribution function of the form

$$C_0(x) = (1-\rho)\sum_{i=0}^{\infty} \rho^i F^{n*}(x),$$

and C

$$C(x) = (1-\rho)\sum_{i=1}^{\infty}\rho^{i-1}F^{n*}(x),$$

where F^{n*} denotes the n-th convolution of F. We have also $F * C_0(x) = C(x)$ and if $F(x)$ is absolutely continuous with respect to Lebesgue measure on R_+, then $C(x)$ is also absolutely continuous, and $C_0(x)$ has an atom at zero furthermore its absolutely continuous part equals C i.e. $C_0(x) = (1-\rho)\delta_0(x) + \rho C(x)$, where δ_0 denotes the unit atom measure at zero. The corresponding Laplace transforms are given by

$$\Psi_0(s) = (1-\rho)/(1-\rho\phi(s)),$$

$$\Psi(s) = (1-\rho)\phi(s)/(1-\rho\phi(s)),$$

where ϕ is the Laplace transform of F.

The corresponding expected values of C_0 and C are $\rho m_F/(1-\rho)$ and $m_F/(1-\rho)$. The following result will be useful in queueing theory.

Theorem B. $C_0 \in NWU$

Proof. [3] Define $S_n = \sum_{i=1}^{n} X_i$, and $M_t = \min\{k : S_k > t\}$, $t > 0$. Since M_t is independent of N_0, it follows from the lack of memory property of the geometric distribution that $(N_0 - M_t \mid N_0 \geq M_t)$ has the same distribution as N_0, and because M_t is a stopping time, $\{X_{M_t} + i, i \geq 1\}$ has the same distribution as $\{X_i, i \geq 1\}$. Thus $(\sum_{M_t+1}^{N_0} X_i \mid N_0 \geq M_t)$ has the same distribution as $\sum_{i=1}^{N_0} X_i = C_0$. The events $\{C_0 > t\}$ and $\{N_0 \geq M_t\}$ are equivalent, thus $I_{C_0>t} = I_{N_0 \geq M_t}$. Now, for $x \geq 0$

$$\mathbf{Pr}(C_0 > t + x) = \mathbf{Pr}((C_0 - t)I_{C_0>t} > x) =$$

$$= \mathbf{Pr}[\{(\sum_{i=1}^{M_t} X_i) - t + \sum_{M_t+1}^{N_0} X_i\}I_{C_0>t} > x]$$

$$\geq \mathbf{Pr}((\sum_{M_t+1}^{N_0} X_i)I_{N_0 \geq M_t} > x) = \mathbf{Pr}(\sum_{M_t+1}^{N_0} X_i > x, N_0 \geq M_t) =$$

$$= \mathbf{Pr}(\sum_{M_t+1}^{N_0} X_i > x \mid N_0 \geq M_t)\mathbf{Pr}(N_0 \geq M_t) = \mathbf{Pr}(C_0 > x)\mathbf{Pr}(N_0 > M_t) =$$

$$= \mathbf{Pr}(C_0 > x)\mathbf{Pr}(C_0 > t).$$

The above inequality, from the definition of the NWU class (see Section 1.5), implies that $C_0 \in NWU$. □

The operation of geometric compounding possesses a number of non banal closure properties which are summarized in the following theorem. Recall that a real function satisfying $(-1)^k f^{(k)}(x) \geq 0$, $k = 0, 1, \ldots$ is called **completely monotonic**.

[3] The proof requires some knowledge about stopping times and conditioning, see Chapter 2 for needed definitions

Theorem C.

(i) *If F has a density which is completely monotonic then C has a completely monotonic density;*

(ii) *If F has a density which log-convex then C has a log-convex density;*

(iii) *If F is DFR then C_0 is DFR;*

(iv) *If F is $HNWUE$ $(HNBUE)$ then C is $HNWUE$ $(HNBUE)$.*

The part (i) was first proved by Keilson (1978), see also Szekli (1987) for a different proof; (ii) was proved by de Bruin and Erdős (1953), (iii) was first proved by Shanthikumar (1988), see also Hansen (1990) for a different approach. The part (iv) is not difficult and is left as an exercise.

Now we turn to the problem how to bound the distance between a geometric compound of nonnegative variables and an exponential distribution with the same mean. This problem is cited by Gertsbakh (1984) as being "of great interest for engineering applications". For the distribution F denote by m_F the first moment, $m_F[2]$ the second moment (if it exists) and let $\gamma_F = m_F[2]/2m_F^2$, $d(C_0, Exp^*) = \sup_{B\in\mathcal{B}} \mid F(B) - Exp^*(B) \mid$, where Exp^* is an exponential distribution with the mean value $m_{C_0} = m_F\rho/(1-\rho)$, $\mathcal{B}$ is the Borel sigma field. The following two theorems are taken from Brown (1990).

Theorem D. *If $F(0) < 1$ and $m_F[2] < \infty$ then*

(i)

$$d(C_0, Exp^*) \le (1-\rho)\max(2\gamma_F, \gamma_F/\rho);$$

(ii)

$$\rho\exp(\frac{-(1-\rho)(2\gamma_F - 1)}{\rho})\exp(\frac{-t}{m_{C_0}}) \le 1 - C_0(t) \le \exp(\frac{-t}{m_{C_0}}) + \gamma_F\frac{1-\rho}{\rho}.$$

Theorem E.

(i) *If F is NBU then $d(C_0, Exp^*) = 1-\rho$;*

(ii) *If F is NWU then $d(C, Exp) \le (2\gamma_F - 1)(1-\rho)$;*

(iii) *If F is DFR then $d(C_0, Exp^*) \le \frac{(1-\rho)\gamma_F}{\gamma_F(1-\rho)+\rho}$;*

(iv) *If $F(0) = \beta$ then $d(C_0, Exp^*) \le \frac{2\gamma_F(1-\rho)(1-\beta)}{1-\beta\rho}$.*

For other bounds and a discussion we refer the reader to Brown (1990).

PROBLEMS AND REMARKS

A. (1) [Wald's Equation] If $X_1, X_2, \ldots$ are i.i.d. random variables having finite expectations, and if N is an integer valued random variable such that $\{N = n\}$ is independent of $X_{n+1}, X_{n+2}, \ldots$ for all $n = 1, 2, \ldots$ and $EN < \infty$, then $E(\sum_{i=1}^{N} X_n) = ENEX$.

(2) $E(C_0) = \rho m_F/(1-\rho)$, $EC = m_F/(1-\rho)$.

B. If F is $HNBUE(HNWUE)$ then C is $HNBUE(HNWUE)$.

C. (1) For a fixed ρ, if we define $B(\rho, m_F, m_F[2])$ to be the best bound for $d(C_0, Exp^*)$ among all distributions with fixed $(m_F, m_F[2])$ and define $B(\rho, \gamma_F)$ to be the best bound for $d(C_0, Exp^*)$ among all distributions with fixed γ_F, then $B(\rho, m_F, m_F[2]) = B(\rho, \gamma_F)$. [Brown (1990)]

(2) $lim_{\rho\to 1}(B(\rho, \gamma_F)/(1-\rho)) = 2\gamma_F$.

D. The simplest shock model assumes that each shock causes a random damage, that damages $\{X_k\}$ on successive shocks are independent and identically distributed, and that failure occurs when the accumulated damage exceeds a specified threshold x. If $G(x)$ is the distribution function of damage sustained from every given shock then the probability that the accumulated damage until the time t, $\sum_{k=0}^{N(t)} X_k$ does not exceed x, is given by
$H(t) = \sum_{k=0}^{\infty} \mathbf{Pr}(N(t) = k)G^{k*}(x)$, where $N(t)$ denotes the number of shocks before time t.

(1) Let $P_k = G^{k*}(x)$, for a fixed x, and assume that $\mathbf{Pr}(N(t) = k) = \frac{(\lambda t)^k}{k!} exp(-\lambda t)$ (Poisson shock model). If (P_k/P_{k-1}) is increasing (decreasing) in k then $1 - H(t)$ is a DFR (IFR) distribution function. [Esary et al. (1973)]

(2) For a Poisson model if G is DFR then the expected time to failure is a concave function of x. [Szekli (1990)]

1.8 Integral orderings for queues

Consider a service node with a fixed number of servers. Customers arrive at the service node, each with some service demand. The node is equipped with a waiting room for not immediately served customers. The capacity of the waiting space may be finite or infinite. If a finite waiting room is occupied entirely the arriving customers are lost. A rule for choosing the next customer for service is called a **queueing discipline**. The most common disciplines are the first-come-first served (FCFS) rule, which selects the customer that arrived earliest of all customers in the queue, and the last-in-first-out (LIFO) rule, which selects the latest arrived customer for service. The processor sharing (PS) discipline allows each arriving customer to receive service at a reduced speed, i.e. the service capacity is distributed equally among all customers present in the node.

What makes queueing theory interesting and useful is the assumption that the sequence of consecutive service times and the sequence of interarrival times are random. A service node with its attendant service times, waiting space and queueing discipline is called a **single node queue.**

A single node queue is described by the symbol $GI/GI/s/L$ which denotes that the interarrival times sequence consists of a collection of general, identically distributed, independent (GI) random variables (renewal arrivals); the sequence of service times is independent, identically distributed (i.i.d) and does not depend on the arrival mechanism. The letter L denotes the capacity of the waiting room (not including the number of servers), s is the number of servers. If $L = \infty$, it is omitted in this notation.

Much of queueing theory is taken up with more restrictive assumptions on interarrival and service times, for example one adopts the assumption that these sequences are i.i.d exponentially distributed. The arrival and service processes in this case are called Poisson processes. The corresponding notation is then $M/M/s/L$.

The most important derived processes of interest for single node systems are the following processes:

- **the queue length** process $(X(t), t > 0)$, where $X(t)$ is the number of customers in the node at the time instant t;
- **the waiting time** process $(W_n, n \in N)$, where W_n is the waiting time for service of the n-th coming customer;
- **the sojourn time** process $(D_n, n \in N)$, where D_n is the time spent by the n-th customer in the system;
- **the departure** process $(N^d(t), t > 0)$, where $N^d(t)$ is the number of customers that completed their service or were lost (overflowing) before the time instant t.
- **busy period** process $(B_n, n \in N)$, where B_n is the length of the n-th consecutive time interval when the queue is not empty.

The idea of a queue forming as a consequence of the interplay of two renewal processes, i.e., the arrival process and the service time process (the $GI/GI/\cdot$ queues) is the basis for nearly all queueing theory. Indeed, this model has served well for nearly 75 years in the major areas of its applications such as telephony and computer performance evaluation. Classical references in the field of single sever queues are for example Cohen (1982), Prabhu (1965), Schassberger (1973), Franken et al. (1981).

There exists a huge literature on stochastic ordering, dealing with theory as well as applications to classical queueing models. The first author who proved monotonicity properties of the nonstationary single node systems was probably Gaede (1965). Whitt (1981) proposes a number of definitions of stochastic ordering, establishes interconnections between these definitions, and derives comparisons for waiting time and queue length in multiserver queues. The book by Stoyan (1983) gives a comprehensive treatment of comparison methods, with applications to multiserver queues up to early eighties. This field is still being under development (see e.g. Baccelli and Bremaud (1993)). We select some classical results and some not available in a book form.

The actual waiting time process $\{W_n\}_{n\geq 0}$ in a $GI/GI/1$ queue with the FCFS discipline is a sequence of random variables satisfying the following recursion formula (see Loynes (1962)).

$$W_{n+1} = \max(0, W_n + S_n - T_n), \ n \geq 0 \tag{1}$$

where W_0 is given as a initial waiting time, $\{S_n\}_{n\geq 0}, \{T_n\}_{n\geq 0}$ are independent i.i.d. sequences of positive random variables. We interpret $\{S_n\}$ as a sequence of service times, and $\{T_n\}$ as a sequence of interarrival times for units arriving to this queueing system.

It is immediate that $\{W_n\}_{n\geq 0}$ forms a Markov chain (see Section 2.4.). The usual probabilistic description of this chain is given by its transition kernel (Markov kernel, see Section 1.3.)

$$k(x,B) = P(W_{n+1} \in B \mid W_n = x),$$

for $B \in \mathcal{B}^1$, and $x \in R$. If W_0 has a fixed initial distribution P^{W_0} then the distribution of W_1, P^{W_1} say, can be thought as a result of a transformation $\mathbf{T}$ which is uniquely defined through $k(x,B)$ by

$$\mathbf{T}\mu(B) = \int_R k(x,B)d\mu(x),$$

for an arbitrary initial measure μ. We have $P^{W_1} = \mathbf{T}P^{W_0}$, and of course $P^{W_{n+1}} = \mathbf{T}P^{W_n}$. We say that $\mathbf{T}$ is strongly stochastically monotone, or equivalently that the corresponding transition kernel k is strongly stochastically monotone if

$$\mu <_{st} \nu \Rightarrow \mathbf{T}\mu <_{st} \mathbf{T}\nu.$$

Lemma A. *The transformation* $\mathbf{T}$ *for a* $GI/GI/1$ *FIFO queue is strongly stochastically monotone.*

Proof. We first show that for $x_1 \leq x_2, k(x_1,(y,\infty)) \leq k(x_2,(y,\infty))$. Indeed,

$$k(x_1,(y,\infty)) = P(W_{n+1} > y \mid W_n = x_1) =$$

$$P((W_n + S_n - T_n)_+ > y \mid W_n = x_1) = P((x_1 + S_n - T_n)_+ > y) \text{ a.s.}$$

The function $(x_1 + S_n - T_n)_+$ is increasing in x_1, hence

$$(x_1 + S_n - T_n)_+ \leq (x_2 + S_n - T_n)_+ \text{ a.s.}$$

i.e.

$$(x_1 + S_n - T_n)_+ <_{st} (x_2 + S_n - T_n)_+.$$

From the definition of $<_{st}$ we have

$$P((x_1 + S_n - T_n)_+ > y) \leq P((x_2 + S_n - T_n)_+ > y).$$

Finally repeating the former transformations we have

$$k(x_1,(y,\infty)) \leq k(x_2,(y,\infty)),$$

i.e. $k(x,(y,\infty))$ is a nondecreasing function of x. To prove the stochastic monotonicity of $\mathbf{T}$, take arbitrary probability measures $\mu <_{st} \nu$. We have

$$\mathbf{T}\mu((y,\infty)) = \int_R k(x,(y,\infty))d\mu(x) \leq \int_R k(x,(y,\infty))d\nu(x) = \mathbf{T}\nu((y,\infty))$$

since k is nondecreasing and $\mu <_{st} \nu$. The above inequality implies $\mathbf{T}\mu <_{st} \mathbf{T}\nu$. □

The above lemma expresses the strong stochastic monotonicity structure of (1) , which can be used to derive two monotonicity properties for $GI/GI/1$ queues: **internal monotonicity** and **external monotonicity**.

Theorem B. (internal monotonicity) *Consider a $GI/GI/1$ FCFS queue. If $W_0 <_{st} W_1$ then $W_n <_{st} W_{n+1}, n \in N$.*

Proof. Since $\mathbf{T}$ is strongly stochastically monotone and $P^{W_0} <_{st} P^{W_1}$ then $\mathbf{T}P^{W_0} <_{st} \mathbf{T}P^{W_1}$ i.e. $W_1 <_{st} W_2$. Proceeding by induction we have $W_n <_{st} W_{n+1}$ for $n \in$N. □

Theorem C. (external monotonicity) *Consider a $GI/GI/1$ FCFS queue. If this system starts to work with two different initial waiting times W_0, and W_0' such that $W_0 <_{st} W_0'$ then $W_n <_{st} W_n', n \in N$.*

Proof. We have for each initial waiting time $P^{W_1} = \mathbf{T}P^{W_0}$, and $P^{W_1'} = \mathbf{T}P^{W_0'}$. Hence from $P^{W_0} <_{st} P^{W_0'}$ and monotonicity of $\mathbf{T}, P^{W_1} <_{st} P^{W_1'}$. By induction the proof is complete. □

From Theorem B. we see that taking $W_0 = 0$, the sequence of waiting times is strongly stochastically increasing and $\lim_{n\to\infty} W_n = W$ exists almost surely. However to assure finiteness and uniqueness of this limit we have to assume that $ES_1/ET_1 < 1$. This ratio is called the **traffic intensity** and denoted by ρ. The distribution of the limit W of waiting times, if it does exist, is called **stationary waiting time distribution.** The problem of existence of stationary waiting times for $G/G/1$ (and other) queues is resolved in quite a general context of asymptotically stationary and ergodic sequences. Classical references are Lindley (1952) and Loynes (1962). Among many books we mention Cohen (1969), Borovkov (1976). For more recent treatment see Franken, Kőnig, Arndt, Schmidt (1981) and Rolski (1981).

For stationary $GI/GI/1$ FCFS queues we have for example the following result (see Stoyan (1983)).

Theorem D. *Consider two $GI/GI/1$ queues with traffic intensities smaller than 1, and stationary waiting times W, W', with finite means, respectively. If $T_1' <_{icv} T_1$ and $S_1 <_{icx} S_1'$ then $W <_{icx} W'$.*

In the case of exponential interarrival times or service times Rolski and Stoyan (1976) showed that it is possible to get the strong stochastic ordering assuming only the convex ordering in the input stream.

Theorem E. *Consider two $M/GI/1$ queues with traffic intensities smaller than 1, and stationary waiting times W, W', respectively. If $ET_1 \geq ET_1'$, and $ES_1 = ES_1'$, $S_1 <_{cx} S_1'$ then $W <_{st} W'$.*

Proof. The stationary waiting time distribution function in an $M/G/1$ queue with an arrival rate $\lambda = 1/ET$ and a service time distribution function G is (see e.g. Cohen (1969), p.255) a geometric compound of the form

$$W(x) = (1-\rho)\sum_{n=0}^{\infty} \rho^n (\mu \int_0^x (1 - G(u)du)^{n*},$$

where $\mu^{-1} = \int_0^\infty x dG(x)$, $\rho = \lambda/\mu$. Now $S_1 <_{cx} S_1'$ is equivalent to $\int_0^x (1 - G(u)du \geq \int_0^x (1 - G'(u)du$, which combined with the above formula gives $W <_{st} W'$. □

Theorem F. *Consider two $GI/M/1$ queues with traffic intensities smaller than 1, and stationary waiting times W, W', respectively. If $T_1 <_{cx} T_1'$ (ET=ET') and $ES_1 \leq ES_1'$ then $W <_{st} W'$.*

Similar results are known for so called tandem queues. A tandem queue is a number of service facilities in series. Customers arrive according to an i.i.d. sequence of interarrival times. Upon arrival, each customer goes to the first station and requires a random amount of service time. If there are other customers present at the time of his arrival, he joins the end of the queue and waits for service. The order of service of customers is FCFS. After being served at the first station, he goes to the second station. Each station operates in a similar fashion but may provide different types of service in general. Every customer has to go through all stations according to the prefixed order and leaves the system after finishing service at the last station. We denote such a system with n stations in tandem by $GI/G_1/ \to G_2/1 \to \cdots \to G_J/1$, where G_i's are the service time distributions in the consecutive stations. Consider two tandem queues with interarrival distributions F, F', and service time distributions G_i, G_i' (at the i-th position in the tandem), respectively. Niu (1981) showed the following theorem

Theorem G.

(i) *If W_n (W_n') is the total waiting time in the system of the n-th customer, $m_F = m_{F'}$, $F <_{cx} F'$ and $G_i <_{icx} G_i'$ ($i = 1, \ldots, J$) then $EW_n \leq EW_n'$ ($n \geq 1$).*

(ii) *If $F = F'$ and $G_i <_{icx} G_i'$, ($i = 1, \ldots, J$) then $W_n <_{icx} W_n'$ ($n \geq 1$).*

A direct approach to stochastic comparisons of queues via random walks, which allows us to treat more general systems was proposed by Harris and Prabhu (1987). Consider a single server queue with interarrival times $\{T_n\}$ and service times $\{S_n\}$. Let $X_n = S_n - T_n$, and assume that $\{X_n, \ n \geq 1\}$ is a sequence of independent random variables with the corresponding distribution functions $K_n(x)$. The main result in Harris and Prabhu (1987) is the following.

Theorem H. *Consider two $GI/G/1$ systems with the corresponding differences $X_n = S_n - T_n$ and $X_n' = S_n' - T_n'$. Assume for these systems the interarival and service times have identical distributions. If $X_n <_{st} X_n'$ ($n \geq 1$) then*

(i) $W_n <_{st} W_n'$;

(ii) *The total idle period up to the n-th arrival epoch in the first system is stochastically larger than in the second one;*

(iii) *During successive busy periods more customers are served in the second system;*

(iv) *If $S_n <_{st} S_n'$ and $T_n' <_{st} T_n$ then the successive busy periods of the first system are shorter than those of the second system;*

For multiserver systems the result (i) was proved by Jackobs and Schach (1972) in a different way (see also Daley and Moran (1968)).

Weaker comparisons are possible for busy periods in $M/GI/1$ queues. Jean-Marie and Liu (1992) proposed the following.

Theorem I. *Consider two $M/GI/1$ queues with the same arrival intensity $\lambda = 1/ET$ and service time distributions G, G', respectively. If $G <_{icx} G'$ then for the stationary busy period distributions B, B', $B <_{icx} B'$.*

PROBLEMS AND REMARKS

A. (Pollaczek-Khinchin formula) In $M/GI/1$ queue the expected waiting time in equilibrium is $EW = \frac{\lambda(\sigma_G^2+m_G^2)}{2(1-\rho)}$, where $\lambda = 1/ET$, σ_G^2 is the service time variance, ρ is the traffic intensity (ES/ET). [Cohen (1969), p.255]

B. (bounds from internal monotonicity) For $GI/GI/1$ queues if $\int_0^\infty exp(\beta x)dK(x) < \infty$, for some $\beta > 0$ and K the distribution of $S_n - T_n$ then

$$1 - a_1 exp(-\theta t) \le W(t) \le 1 - a_2 exp(-\theta t),\ t \ge 0,$$

where θ is such that $\int_{-\infty}^\infty exp(\theta x)dK(x) \le 1$, $a_1 = sup_{t>0} f(t)$, $a_2 = inf_{t>0} f(t)$, for $f(t) = (1-K(t))/\int_t^\infty exp(\theta(y-t))dK(y)$. [Stoyan (1983), p.84]

C. (bounds from external monotonicity) For $GI/GI/1$ queues in equilibrium

$$EW \ge \frac{\sigma_G^2}{2m_F(1-\rho)} - \frac{1}{2m_G},$$

where $m_F = ET$ is the mean interarrival time. [Stoyan (1983), p.87]

D. (Kingman's bound) For $GI/GI/1$ queues in equilibrium

(1)

$$EW \le \frac{\sigma_F^2 + \sigma_G^2}{2m_F(1-\rho)};$$

[Kingman (1962)];

(2)

$$EW \le \frac{\sigma_F^2\rho(2-\rho) + \sigma_G^2}{2m_F(1-\rho)}.$$

[Daley (1977)].

E. For the stationary busy period B of a $GI/GI/1$ queue with NBU (NWU) interarrivals

$$EB \ge (\le) \frac{ES}{1 - E(M(S))},$$

where $M(t) = \sum_{n=1}^\infty F^{n*}(t)$ is the renewal function of the interarrival distribution function F.

F. Let X be an integer valued random variable. Consider a single server queue with batch arrivals (denoted by $GI^X/GI/1$), with i.i.d. bulk sizes distributed according to X, with i.i.d interarrival times $\{T_n\}$, and i.i.d. service times $\{S_n\}$. Denote by $\{W_n\}$ the sequence of consecutive bulk waiting times ($W_0 = 0$), and by $\{R_n\}$ customer response times. (For stationary characteristics we drop the index n).

(1) In two systems $M^X/GI/1$, and $M^{X'}/GI/1$ with the same arrival rate and the same service times, if $X <_{cx} X'$ (the same mean value) then $W <_{st} W'$ and $R <_{st} R'$. [Chang (1990)] (see also Rolski (1976)).

(2) For $GI^X/GI/1$ and $GI^{X'}/X/1$, if $X_{icx}X'$, $S <_{icx} S'$, $T' <_{cx} T$ then $W <_{icx} W'$ and $R <_{icx} R'$. [Jean-Marie and Liu (1992)].

G. (Little's formula) For a $GI/GI/1$ queue in equilibrium ($\rho < 1$), $\bar{L} = \lambda \bar{W}$, where $\bar{L} = \lim_{t\to\infty}(1/t)\int_0^t L(s)ds$, for $L(s)$ the number of customers waiting in the queue at the moment s, $\lambda = \lim_{n\to\infty}(1/n)\sum_{i=0}^n T_i$ and $\bar{W} = \lim_{n\to\infty}(1/n)\sum_{i=0}^n W_i$. [Stidham (1974)].

1.9 Relative inverse orderings for queues

In this section the results on geometric compounding from Section 1.7 will be applied in the context of queueing theory. Recall (see Section 1.5) that for example $F \in DFR \Leftrightarrow Exp <_c F$ (and likewise for other classes, IFR, NWU, NBU, etc.), so the closure properties for the DFR class of distributions can be interpreted as some closure properties of the corresponding ordering $<_c$, (similarly for $<_{su}$ and other relative inverse orderings).

Consider a $GI/GI/1$ queue with the interarrival times $\{T_n,\ n \geq 1\}$, and service times $\{S_n,\ n \geq 1\}$. Assume that $\rho = ES/ET < 1$. The main link between compounding and queueing theory is the following well known theorem (see Feller (1971)).

Theorem A. *The stationary waiting time distribution $W(x)$ in $GI/GI/1$ systems is a geometric compound of the following form*

$$W(x) = \sum_{i=0}^{\infty} \alpha(1-\alpha)^i H^{i*}(x),$$

where $\alpha = W(0+)$, $H(x)$ is the normalized distribution function of the first ascending ladder hight in the random walk $\{S_n - T_n\}$.

From the above theorem and Theorem 1.7.B. we have immediately

Theorem B. (Köllerström (1976)) *The stationary waiting time distribution W in $GI/GI/1$ systems is NWU.*

In order to utilize Theorem A. in the case of DFR distributions we use a technical lemma (see Szekli (1986)).

Lemma C. *If the distribution of S_n in the random walk $\{S_n - T_n\}$ is DFR then $H(x)$ is DFR*

Now from Theorem 1.7. C. , and Lemma C. we have

Theorem D. *Consider a $GI/GI/1$ queue in equilibrium.*

(i) *If the service time distribution has a completely monotonic density then the stationary waiting time distribution W has a completely monotonic density;*

(ii) *If the service time distribution has a logconvex density then the stationary waiting time distribution W has a logconvex density;*

(iii) *If the service time distribution is DFR then W is DFR;*

(iv) *If the service time distribution is DFR and interarrival times are exponentially distributed then W has a logconvex density.*

For $GI/GI/k$ systems with $k > 1$ the geometric compound formula from Theorem A. is not available but using alternative methods we have (see Szekli (1987))

Theorem E. *In $GI/GI/k$ ($k = 1, 2, \ldots$) systems if the service time distribution has a completely monotonic density then the stationary waiting distribution W has a completely monotonic density.*

The above theorem for $M/GI/1$ systems was first proved by Keilson (1978). He also gave the following result

Theorem F. *The busy period distribution function in $M/GI/1$ queues has a completely monotonic density provided the service time distribution has a completely monotonic density.*

Geometric compounding comes in view in queueing theory also in a relationship with busy periods in $M/GI/\infty$ systems. This observation is based on the following.

Theorem G. (Stadje (1985)) *The busy period distribution $B(x)$ in $M/GI/\infty$ queue is given by*

$$B(x) = 1 - \frac{1}{\lambda} \sum_{k=1}^{\infty} d^{n*}(x),$$

where $d(x) = \frac{dD(x)}{dx}$, $D(x) = 1 - exp(-\lambda \int_0^x 1 - G(t)dt)$, G is the service time distribution function and λ is the arrival intensity.

Corollary H.

(i) *If the mean service time ES in $M/GI/\infty$ queue is finite then $1 - B(x)$ equals to a scaled density of a geometric compound i.e.*

$$1 - B(x) = a_1(1-\beta) \sum_{k=1}^{\infty} \beta^{k-1} \check{d}^{k*}(x),$$

where $\beta = 1 - exp(-\lambda ES)$, $\check{d}(x) = d(x)/\beta$, and $a = \frac{\beta}{\lambda(1-\beta)}$.

(ii) *If the mean service time ES in $M/GI/\infty$ queue is infinite then $1 - B(x)$ equals to a scaled renewal density*

$$1 - B(x) = a_2 \sum_{k=1}^{\infty} d^{k*}(x),$$

where $a_2 = \frac{1}{\lambda}$.

Now we have

Theorem I.

(i) *If the service time distribution in a $M/GI/\infty$ queue has a completely monotonic density then $B(x)$ has a complete monotonic density.*

(ii) *If the service time distribution in a $M/GI/\infty$ queue is DFR and $ES < \infty$ then the busy period distribution B is DFR.*

Proof. (*i*) Since

$$d(x) = \lambda(1 - G(x))exp(-\lambda \int_0^x 1 - G(t)dt),$$

and functions of the form $exp(-\Psi)$, with Ψ having completely monotonic derivative, are completely monotonic (see Feller (1971)) hence d is completely monotonic as a product of two completely monotonic functions. If the mean service time is finite then from the above corollary and Theorem 1.7.C. (i) we conclude that $1 - B(x)$ is a completely monotonic function, which implies the complete monotonicity of its density. The renewal density of a distribution with a complete monotonic density is completely monotonic (see Keilson (1978)) therefore in the case when ES is infinite we get from the above corollary the complete monotonicity of $1 - B(x)$.
(ii) In this case d is logconvex as a product of two logconvex functions. From Theorem 1.7.C. it follows that $1 - B(x)$ is a logconvex function, which is equivalent to $B \in DFR$ (see Section 1.5). □

PROBLEMS AND REMARKS

A. Consider a $GI/GI/1$ queue in equilibrium. If the interarrival distribution function F is DFR then the idle period distribution function I is DFR. [Szekli (1986)].

B. The number of customers served during a busy period of an $M/GI/1$ queue with IFR service times is DFR. [Shanthikumar (1988)].

1.10 Loss systems

One of the basic models in telephony is so called loss system, i.e. a queueing system with finite waiting room, which we denote according to the notation introduced in Section 1.8 by $G/G/s/L$. If $L = 0$ then it is called a pure loss system. Such a system operates under $FCFS$ discipline and when incoming customers (calls) find the system full they are lost in the sense that they are not waiting for service nor they come back into the system afterwards. If interarrival distances and service times form i.i.d. sequences, which are independent then we use $GI/GI/s/L$ notation, and if in addition the distribution function of interarrival times $F(x)$ and the distribution of service times $G(x)$ are exponential then the system is denoted by $M/M/s/L$.

A central role in the study of loss systems plays the process $\{X(t)\}$ of the number of busy servers (lines), indexed by time t. Various technics from the theory of Markov processes have been used to study a long run behavior of this process, based on so called ergodic theorems (see Section 2.4), and embedded Markov chains. Especially, loss probabilities have been of interest while studying stationary (long run) properties of the system.

There are at least two points of view on the specification of the probability of loss. First, it is the probability of the loss of a call when operating under stationary conditions. More precisely, this is an operator (server) point of view, one is interested in a limiting behavior of $\mathbf{Pr}(X(t) = s)$ with $t \to \infty$, where s is the number of available servers (lines) of the system. Usually the existence and form of $p_s = \lim_{t\to\infty} \mathbf{Pr}(X(t) = s)$ is studied. Secondly, it is stationary probability that an arriving customer (call) finds all servers (lines) busy and therefore is lost. This is a customer point of view, one is interested in a limiting behavior of $\mathbf{Pr}(X(\tau_n-) = s)$, where τ_n's denote the times of consecutive arrivals of customers (calls) into the system (τ_n- means just before the arrival). In general, the above limiting values (if they do exist) need not be equal. Moreover, the stochastic processes $\{X(t)\}$ and $\{X(\tau_n-)\}$ can not be simultaneously strictly stationary on the same probability space, the time stationarity of $\{X(t)\}$ precludes the stationarity of the sequence $\{X(\tau_n-)\}$ and vice versa (see e.g. Franken et al. (1981)). However, in the classical case of Poisson arrivals, i.e. when the interarrival distances form an i.i.d. sequence exponentially distributed, the customer and server (time) characteristics coincide. Such coincidence is called in the literature the PASTA property, for Poisson arrivals see time averages (see e.g. Wolff (1982)).

The formula for the loss probability in $M/GI/s/0$ is well known as Erlang's loss formula

$$p_s = \frac{\frac{(\lambda\mu)^s}{s!}}{\sum_{i=0}^{s} \frac{(\lambda\mu)^i}{i!}}.$$

This formula has an interesting history. Erlang (1917) established it with service times deterministic (constant) but the proof was incomplete, in the case of exponential service times his proof was acceptable. These investigations raised the problem, can the service time be arbitrary? The positive answer was given by Valout and in subsequent time by other researchers to mention only Pollaczek, Palm, Kosten, Khinchin, Sevastianov. The most satisfactory in a mathematical sense is the proof by Sevastianov (1957), based on an ergodic theorem for Markov processes (see Section 2.4.), which is of interest also in other context. Sevastianov used the variation distance v between two arbitrary distributions P, Q on a measurable space $(\Omega, \mathcal{F})$, $v(P,Q) = \int_\Omega | P(d\omega) - Q(d\omega) |$. If a family (P_t) of distributions is indexed by $t \geq 0$ then this family converges in variation to a limit Q, $P_t \to^v Q$ if $\lim_{t\to\infty} v(P_t, Q) = 0$. He proved for $M/GI/s/0$ systems that the distribution of an enlarged process $(X(t), S_t^1, \ldots, S_t^{X(t)})$, where S^j's denote the residual service times in the occupied lines at time t, converges in variation to a limiting distribution, which is independent of the initial conditions of the system. Immediately from his Theorem 3, it follows

Theorem A. *In a $M/GI/s/0$ queue*

$$\lim_{t\to\infty} \mathbf{Pr}(X(t) = k) = \frac{\frac{(\lambda\mu)^k}{k!}}{\sum_{i=0}^{s} \frac{(\lambda\mu)^i}{i!}}, \quad k = 1, \ldots, s.$$

independently on the initial conditions, where λ is the arrival intensity, μ is the mean service time.

From a more direct analysis, Takacs (1969) obtained limiting probabilities for the enlarged process taken at the arrival times $(X(\tau_n-), S^1_{\tau_n}, \ldots, S^{X(\tau_n-)}_{\tau_n})$. As a corollary we have

Theorem B. *In a $M/GI/s/0$ queue*

$$\lim_{n\to\infty} \mathbf{Pr}(X(\tau_n-) = k) = \frac{\frac{(\lambda\mu)^k}{k!}}{\sum_{i=0}^{s} \frac{(\lambda\mu)^i}{i!}}, \quad k = 1, \ldots, s.$$

Another class of tractable queueing loss systems consists of $M/GI/1/L$ systems with a finite waiting room, where the two dimensional process of the number of customers in the system and the remaining service time of the customer in service (or related processes obtained by embedding at arrival or departure epochs) is the typical starting point of classical analysis, see e.g. Cohen (1982), Keilson (1966), Kendall (1953), Takacs (1969). A transform-free analysis of $M/GI/1/L$ queues is proposed by Niu and Cooper (1991). However, for general interarrival times and service times stochastic comparison results are rare. For pure loss systems we have

Theorem C. *Consider two $GI/GI/1/0$ queues in equilibrium, with interarrival distances $\{T_n\}$, service times $\{S_n\}$ and $\{T'_n\}$, $\{S'_n\}$, respectively.*

(i) *If $T' <_{st} T$ and $B <_{st} B'$ then for the stationary probabilities that an arriving customer finds all servers busy and is lost $p_{loss} \le p'_{loss}$;*

(ii) *If the interarrival times to both queues have the same distribution which is DFR and $B <_{icv} B'$ then $p_{loss} \le p'_{loss}$;*

(iii) *If the interarrival times to both queues have the same distribution which is DFR and $B' <_{cx} B$ (the same mean values) then $p_{loss} \le p'_{loss}$.*

Proof. For systems of the type $GI/GI/1/0$ it can be checked that

$$p_{loss} = 1 - \frac{1}{\int_0^\infty U(t)dG(t)},$$

where $U(t) = \sum_{i=0}^{\infty} F^{n*}(t)$, and F, G are the interarrival and service time distribution functions, respectively. Now, (i) is immediate, (ii) follows from the fact that the renewal function $U(t)$ for F which is DFR is an increasing concave function of t (see Section 2.10), and (iii) is in addition a consequence of the fact that $B <_{icv} B'$ is equivalent to $B' <_{cx} B$ for the same mean values (see Section 1.3). □

An intuitive meaning of (iii) is remarkable, a queue with a fixed traffic intensity can have a greater loss probability if we make the service time less variable. This fact should be compared with the following results of Miyazawa (1989), where for Poisson arrival streams a reversed direction monotonicity is present, i.e. more variable service times cause greater loss probabilities.

Theorem D. *Consider two $M/GI/1/L$ systems in equilibrium with the same arrival rate and the service time distributions G, G', respectively. If $G <_{cx} G'$ (the same mean values), $\rho < 1$, then $p_{loss} \le p'_{loss}$.*

Miyazawa (1989, 1990) proposed also results for batch arrival queues.

Theorem E. *Consider two $GI^X/GI/1/L$ systems in equilibrium with the same traffic intensities*

(i) *If F and G are both NBU, F' and G' are both exponential, with the same mean values as F, G, respectively, and the batch sizes have the same distribution then $p_{loss} \le p'_{loss}$;*

(ii) *If $F = F'$ is exponential, $G = G'$ is exponential (the systems are of type $M^X/M/1/L$), and if for the batch sizes we have $X <_{cx} X'$ then $p_{loss} \le p'_{loss}$;*

(iii) *If $F = F'$ is exponential, $G <_{cx} G'$ (the same mean values) then $p_{loss} \le p'_{loss}$.*

Miyazawa (1989) conjectured that (ii) in the above theorem remains true for general not necessarily exponential distributions. (He formulated (i) for a broader class $NBUE$).

PROBLEMS AND REMARKS

A. For $GI/M/1/0$ systems if $F <_{cx} F'$ then $p_{loss} \le p'_{loss}$. (For an extension to ergodically stable inputs see Chang and Pinedo (1989)).

B. Let p^n_{loss} be the loss probability in $M/E_n/1/L$ system, where E_n denotes the n-th order Erlang distribution with a fixed mean value μ for all n. The sequence p^n_{loss} is nonincreasing in n. [Miyazawa (1989)]

C. (Proportional queue length distributions) For $M/GI/1/L$ queues in equilibrium

$$\frac{\mathbf{Pr}(X_L(t) = n)}{\mathbf{Pr}(X_L(t) = m)} = \frac{\mathbf{Pr}(X_{L'}(t) = n)}{\mathbf{Pr}(X_{L'}(t) = m)},$$

where $X_L(t)$ denotes the corresponding queue length, $0 < L < L'$, $m, n \le L$. [Keilson (1966), Glasserman and Gong (1991)]

D. Consider a $M/G/2/0$ system with an ordered entry, i.e. the servers in the system are numbered 1, 2, and the service time of the server 1 is E_1, exponential with the mean value μ^{-1}, the service time of the server 2 is E_2, Erlang of the 2-nd order with the mean value μ^{-1}. If each arriving customer is assigned to the lowest number idle server the corresponding loss probability we denote by $p_{(E_1,E_2)}$, and if the order is reversed by $p_{(E_2,E_1)}$. These loss probabilities are related by $p_{(E_2,E_1)} < p_{(E_1,E_1)} = p_{(E_2,E_2)} < p_{(E_1,E_2)}$. [Cooper and Palakurthi (1989)].

E. In $M/GI/s/L$ systems

$$(1 - \frac{1}{\rho}) \le p_{loss} \le 1 - \frac{1}{\rho + 1},$$

where $\rho = \lambda/(s\mu)$, for the corresponding arrival rate λ and service rate μ. [Sobel (1980)].

F. Consider two $G/M/2/0$ systems with equally distributed service times and the interarival distances $\{T_n\}$, $\{T_n'\}$, which form stationary sequences. If $Ef(T_1,\ldots,T_n) \leq Ef(T_1',\ldots,T_n')$ for all convex functions f and all $n \geq 1$ then $p_{loss} \leq p_{loss}'$. [Fleischmann (1976)].

G. The interdeparture times from $M/GI/s/0$ systems (including served and lost customers) form i.i.d exponentially distributed sequence, i.e. the departure process is Poisson. [Shanbhag and Tambouratzis (1973)].

H. In the class of $GI/M/s/0$ systems with a given mean interarrival time, the lost probability p_{loss} is least in $D/M/s/0$ (D denotes deterministic i.e. constant nonrandom).[Beneš (1959)]. It is an old conjecture that this result is true for systems with a stationary arrival stream.

Chapter 2

Multivariate Ordering

We have presented in Chapter 1 a collection of univariate orderings which are commonly used in the literature. There is also a large collection of orderings used in more dimensional spaces. For a general treatment we refer the reader to books by Marshall and Olkin (1979), Tong (1980), Dharmadhikari and Joag-Dev (1988), Shaked and Shanthikumar (1994), and references therein. In this chapter we concentrate ourselves on multivariate strong stochastic orderings because, it is possible to provide for them some coupling constructions, and they still play a dominant role in applications. We start with Strassen's theorem which is in a sense fundamental for strong stochastic orderings. This theorem asserts the existence of almost surely comparable versions of two vectors which are strongly stochastically ordered. Unfortunately the proof is carried out in a not effective way, not giving us a method how to construct such vectors. Such constructions are possible under some sufficient conditions, and we provide in this chapter two detailed *coupling constructions* of almost surely comparable random vectors which lead naturally to strong stochastic ordering. These constructions give us a method for proving strong stochastic ordering in applied probability models. A similar approach will be used then to point processes.

2.1 Strassen's theorem

For $\mathbf{x} = (x_1, \ldots, x_n), \mathbf{y} = (y_1, \ldots, y_n)$ we define the coordinatewise ordering by $\mathbf{x} \leq \mathbf{y}$ if $x_i \leq y_i, i = 1, \ldots, n$. A real function $f : R^n \to R$ is nondecreasing (nonincreasing) if $\mathbf{x} \leq \mathbf{y}$ implies $f(\mathbf{x}) \leq f(\mathbf{y})$ for all $\mathbf{x}, \mathbf{y} \in R^n$.

Random vectors $\mathbf{X} = (X_1, \ldots, X_n)$ and $\mathbf{Y} = (Y_1, \ldots, Y_n)$ are **strongly stochastically** ordered, in symbols $\mathbf{X} <_{st} \mathbf{Y}$, if

$$Ef(\mathbf{X}) \leq Ef(\mathbf{Y}),$$

for all nondecreasing functions f for which the expectations exist.

If $\mathbf{X} <_{st} \mathbf{Y}$ then we also write $P^{\mathbf{X}} <_{st} P^{\mathbf{Y}}$, for the corresponding distributions, which are probability measures on $\mathcal{B}^n$.

The following theorem follows from results of Strassen (1965). A simpler proof can be found in Liggett's (1985) book. We give a sketch of a measure theoretical proof.

Theorem A. *For two random vectors* $\mathbf{X} <_{st} \mathbf{Y}$ *if and only if there exist versions* $\check{\mathbf{X}}, \check{\mathbf{Y}}$ *of* $\mathbf{X}, \mathbf{Y}$ *on a common probability space, such that* $\check{\mathbf{X}} \leq \check{\mathbf{Y}}$ *a.s.*

Proof. The "if " part of the theorem is obvious. To prove the "only if" part of the theorem we need two following lemmas.

Lemma B. (Guy (1961), Kelley (1959)) *If α and β are finitely additive probability measures on fields $\mathcal{A}$, $\mathcal{B}$, respectively, and $\mathcal{C}$ is a field containing both $\mathcal{A}$ and $\mathcal{B}$ then the following conditions are equivalent*

(i) *there exist a finite additive probability measure P on $\mathcal{C}$ which is a common extension of μ and ν;*

(ii) *If $A \subseteq B$, $A \in \mathcal{A}, B \in \mathcal{B}$ then $\alpha(A) \leq \beta(B)$.*

Lemma C. *If μ and ν are probability measures on a Polish space E and $D \subseteq E \times E$ is closed then the following conditions are equivalent*

(i) *there exist a probability measure P on $E \times E$ such that $P(D) = 1$ and μ, ν are marginals of P;*

(ii) *for Borel sets A, B*

$$(A \times B) \cap D = \emptyset \Rightarrow \mu(A) + \nu(B) \leq 1.$$

Proof of Lemma C. . (i)→(ii). Let $(A \times B) \cap D = \emptyset$, we have

$$\mu(A) = P(A \times E) = P((A \times E) \cap D) = P((A \times (E \setminus B)) \cap D) \leq$$

$$P(E \times (E \setminus B)) = \nu(E \setminus B) = 1 - \nu(B).$$

(ii)→(i). Let $\mathcal{A} = \{(A \times E) \cap D : A \text{ } Borel\}$, $\mathcal{B} = \{(E \times B) \cap D : B \text{ } Borel\}$, $\alpha((A \times E) \cap D) = \mu(A)$, $\beta((E \times B) \cap D) = \nu(B)$. Suppose $(A \times E) \cap D \subseteq (E \times B) \cap D$, then $(A \times (E \setminus B)) \cap D = \emptyset$, so $\mu(A) + \nu(E \setminus B) \leq 1$, which implies $\alpha((A \times E) \cap D) \leq \beta((E \times B) \cap D)$. We use now Lemma B. to prove the existence of P, and the proof of Lemma C. is completed.

In order to continue the **proof** of Theorem A. , let $D = \{(\mathbf{x}, \mathbf{y}) : \mathbf{x} \leq \mathbf{y}\}$, $\mathbf{x}, \mathbf{y} \in R^n$, and A, B be such that $(A \times B) \cap D = \emptyset$. By $\vec{A}$ denote the minimal increasing set containing A, i.e. $\vec{A} = \{\mathbf{x} : \mathbf{a} \leq \mathbf{x},\ \mathbf{a} \in A\}$. Now from the assumption that $\mathbf{X} <_{st} \mathbf{Y}$ we have

$$P^{\mathbf{X}}(A) \leq P^{\mathbf{X}}(\vec{A}) \leq P^{\mathbf{Y}}(\vec{A}) \leq P^{\mathbf{Y}}(R^n \setminus B) = 1 - P^{\mathbf{Y}}(B)$$

because $\vec{A} \subseteq R^n \setminus B$, indeed if $\vec{A} \cap B$ were not empty then $\mathbf{x} \in \vec{A} \cap B$ would imply that $(\mathbf{a}, \mathbf{x}) \in (A \times B) \cap D$ for some $\mathbf{a} \leq \mathbf{x}$, contradicting $(A \times B) \cap D = \emptyset$.
Now, since (ii) of Lemma C. is valid for $P^{\mathbf{X}}$, $P^{\mathbf{Y}}$, it follows that there exist P with marginals $P^{\mathbf{X}}$, $P^{\mathbf{Y}}$, with $P(D) = 1$. Taking for example the canonical space representation, we obtain versions $\check{\mathbf{X}}, \check{\mathbf{Y}}$ of $\mathbf{X}, \mathbf{Y}$ on a common probability space, such that $\check{\mathbf{X}} \leq \check{\mathbf{Y}}$ a.s. □

Since Lemma C. is true for arbitrary Polish spaces, random vectors $\mathbf{X}$, $\mathbf{Y}$ can be taken to have values in an arbitrary partially ordered Polish space, with a closed partial order.
Another ordering than the coordinatewise ordering, which is used to define a strong stochastic ordering in R^n is the majorization.

Majorization

For two vectors $\mathbf{x}, \mathbf{y} \in R^n$ we define the relation $\mathbf{x} \prec \mathbf{y}$ by $\sum_{i=1}^{k} x_{[i]} \leq \sum_{i=1}^{k} y_{[i]}$, $k < n$, $\sum_{i=1}^{n} x_{[i]} = \sum_{i=1}^{n} y_{[i]}$, where $x_{[1]} \geq \ldots \geq x_{[n]}$ denotes nonincreasing rearrangement of $\mathbf{x}$. We call this relation the **majorization.**

The majorization is not a partial order on the whole R^n, but it is a partial order for example on $D = \{\mathbf{x} \in R^n : x_1 \geq \ldots \geq x_n\}$. Therefore it is reasonable to consider a strong stochastic ordering generated by this relation. The corresponding Strassen's type theorem are known and can be found e.g. in Marshall and Olkin (1979), Chapter 11 A.

We recall some characterizations of this relation. The following lemma suggests that many properties of majorization can be reduced into two dimensional case.

Lemma D. *If* $\mathbf{x} \prec \mathbf{y}$ *then there exists a sequence* $\mathbf{z}^1, \ldots, \mathbf{z}^k$ *such that*

$$\mathbf{x} \prec \mathbf{z}^1 \prec \ldots \prec \mathbf{z}^k \prec \mathbf{y},$$

and each two vectors $\mathbf{z}^i, \mathbf{z}^{i+1}, i = 1, \ldots, n-1$ *have only two different coordinates.*

Proof. Without loss of generality we can assume that $x_1 \geq \ldots \geq x_n, y_1 \geq \ldots \geq y_n$. Let $\mathbf{z}^1$=$\mathbf{x}$. We have

$$z_1^1 + \ldots + z_{n-1}^1 \leq y_1 + \ldots + y_{n-1},$$
$$z_1^1 + \ldots + z_{n-1}^1 + z_n^1 = y_1 + \ldots + y_{n-1} + y_n,$$

hence $z_n^1 \geq y_n$.
Define

$$z_i^2 = z_i^1, i = 1, \ldots, n-2,$$
$$z_{n-1}^2 = z_{n-1}^1 - y_n,$$
$$z_n^2 = y_n.$$

Since

$$z_{n-1}^1 + z_n^1 = z_{n-1}^2 + z_n^2,$$

we have

$$z_1^1 + \ldots + z_n^1 = z_1^2 + \ldots + z_n^2 = y_1 + \ldots + y_n.$$

Now $z_{n-1}^2 \geq z_{n-1}^1, z_n^2 = y_n$, so

$$z_1^1 + \ldots z_{n-1}^1 \leq z_1^2 + \ldots + z_{n-1}^2 = y_1 + \ldots + y_{n-1},$$

i.e. $\mathbf{z}^1 \prec \mathbf{z}^2 \prec \mathbf{y}$.
Applying the same transformation to $(z_1^2, \ldots, z_{n-1}^2), (y_1, \ldots, y_{n-1})$ we obtain $\mathbf{z}^3$ with $z_{n-1}^3 = y_{n-1}$, so by induction we can complete the proof. □

The following characterization was given by Hardy, Littlewood and Polya (1952). For other equivalent conditions see e.g. Rüschendorf (1981).

Theorem E. *We have* $\mathbf{x} \prec \mathbf{y}$ *if and only if*

$$\sum_{i=1}^{n} g(x_i) \leq \sum_{i=1}^{n} g(y_i), \tag{1}$$

for all real convex functions g.

Proof. From Lemma D. it is enough to consider $n = 2$. Let $(x_1, x_2) \prec (y_1, y_2)$ and g be convex. Assume $y_1 \neq y_2$ (otherwise $(x_1, y_1) = (x_2, y_2)$) then $x_1 = \alpha y_1 + (1-\alpha)y_2$, $x_2 = \alpha y_2 + (1-\alpha)y_1$, for $\alpha = \frac{x_1 - y_2}{y_1 - y_2} \in [0,1]$. Thus $g(x_1)+g(x_2) = g(\alpha y_1+(1-\alpha)y_2)+g(\alpha y_2+(1-\alpha)y_1) \leq g(y_1)+g(y_2)$. Now assume (1), and take $g(z) = \max(z - y_{[k]}, 0)$ for $k \leq n$. Since g is convex we apply (1), which immediately gives $\mathbf{x} \prec \mathbf{y}$. □

A function $h : R^n \to R$ such that $\mathbf{x} \prec \mathbf{y}$ implies $h(\mathbf{x}) \leq (\geq) h(\mathbf{y})$ is called **Schur convex (concave)**.

Just as ordinary convexity can be characterized by a monotonicity property of the derivative, so also can Schur convexity (see e.g. Roberts and Varberg (1973)).

Theorem F. *If f is symmetric and has continuous partial derivatives on R^n then f is Schur convex if and only if $(x_2 - x_1)(\partial f/\partial x_2 - \partial f/\partial x_1) \geq 0$.*

A Schur convex function need not be convex ($f(x_1, x_2) = | x_2 - x_1 |^{1/2}$) and convex function need not be Schur convex ($f(x_1, x_2) = x_1 + x_2^2$). However, we have the following result (see e.g. Roberts and Varberg (1973))

Theorem G. *A convex function f is Schur convex if and only if it is symmetric.*

PROBLEMS AND REMARKS

A. Let $\mathbf{X} = (X_1, \ldots, X_n)$, and $\mathbf{Y} = (Y_1, \ldots, Y_n)$ be two random vectors. $\mathbf{X} <_{st} \mathbf{Y}$ iff $\phi(\mathbf{X}) <_{st} \phi(\mathbf{Y})$, for all increasing functions $\phi : R^n \to R$.

B. Let $\mathbf{X} = (X_1, \ldots, X_n)$, and $\mathbf{Y} = (Y_1, \ldots, Y_n)$ be two random vectors. $\mathbf{X} <_{st} \mathbf{Y}$ iff $\mathbf{Pr}(\mathbf{X} \in A) \leq \mathbf{Pr}(\mathbf{Y} \in A)$, for all measurable sets $A \subseteq R^n$ with increasing indicator functions.

C. If $\mathbf{X} <_{st} \mathbf{Y}$ then $\mathbf{Pr}(X_1 > t_1, \ldots, X_n > t_n) \leq \mathbf{Pr}(Y_1 > t_1, \ldots, Y_n > t_n)$, for all t_i's.

D. If $\mathbf{X} <_{st} \mathbf{Y}$ then $\mathbf{Pr}(X_1 \leq t_1, \ldots, X_n \leq t_n) \geq \mathbf{Pr}(Y_1 \leq t_1, \ldots, Y_n \leq t_n)$, for all t_i's.

E. Let $X_1, \ldots, X_n$ be independent with logconcave densities then $(\mathbf{X} \mid \sum_{i=1}^n X_i = s) <_{st} (\mathbf{X} \mid \sum_{i=1}^n X_i = t)$, for $s \leq t$.[Efron (1965), Shanthikumar (1987), Daduna and Szekli (1992)].

F. Let $\mathbf{X} = (X_1, \ldots, X_n)$, and $\mathbf{Y} = (Y_1, \ldots, Y_n)$ be two random vectors. Define $\mathbf{X} <_{wk} \mathbf{Y}$ iff $\mathbf{Pr}(\mathbf{X} \in A) \leq \mathbf{Pr}(\mathbf{Y} \in A)$, for all measurable sets A of the form $\{\mathbf{y} : \mathbf{x} \leq \mathbf{y}\}$, $\mathbf{x} \in R^n$ (upper intervals). The following conditions are equivalent

(1) $\mathbf{X} <_{wk} \mathbf{Y}$;

(2) $E(\psi(\mathbf{X})) \leq E(\psi(\mathbf{Y}))$ for all $\psi(x_1, \ldots, x_n) = f_1(x_1) \cdots f_n(x_n)$, where f_i's are increasing, [Bergman (1978)];

(3) $E(\phi(\mathbf{X})) \leq E(\phi(\mathbf{Y}))$ for all ϕ which are Δ-monotone as functions of any k of the n variables. [Rüschendorf (1980)].

G. The following functions are Schur convex:

(1) $\sum_{i=1}^n x_i \ln x_i$, $\mathbf{x} \geq 0$;

(2) $-\sum_{i=1}^n \ln x_i$, $\mathbf{x} > 0$;

(3) $\sum_{i=1}^n \frac{1}{x_i}$, $\mathbf{x} > 0$;

(4) $[\frac{1}{n}\sum_{i=1}^n x_i^2]^{1/2}$, $\mathbf{x} > 0$.

H. If we denote by $\sum! F(a_1, \ldots, a_n)$ the sum of the $n!$ terms obtained from $F(a_1, \ldots, a_n)$ by the possible permutations of the $\mathbf{a}$, the symmetrical mean of $F(\mathbf{a})$ is $(1/n!)\sum! F(\mathbf{a})$. In the special case $F_{[\alpha]}(\mathbf{a}) = a_1^{\alpha_1} \ldots a_n^{\alpha_n}, a_i > 0, \alpha = (\alpha_1, \ldots, \alpha_n), \alpha_i \geq 0, i = 1, \ldots, n$, taking $\alpha = [1, 0, \ldots, 0]$ we obtain the arithmetic mean, with $\alpha = [1/n, \ldots, 1/n]$ we get the geometric mean. If $\alpha \prec \alpha'$ then $(1/n!)\sum! F_{[\alpha]}(\mathbf{a}) \leq (1/n!)\sum! F_{[\alpha']}(\mathbf{a})$, for all $\mathbf{a} > 0$, i.e. the symmetrical means are Schur convex functions of α. Furthermore , $a_1^{1/n} \ldots a_n^{1/n} < (1/n!)\sum! F_{[\alpha]}(\mathbf{a}) < (a_1 + \ldots + a_n)/n$, if $\alpha_1 + \ldots + \alpha_n = 1$, and $\mathbf{a}$ has not constant coordinates.

2.2 Coupling constructions

From Strassen's theorem we know that for two strongly stochastically ordered random vectors there exist versions which are a.s. comparable. However it is not in general evident how to construct such versions. On the other hand if we can construct two random vectors a.s. comparable, such a construction assures the strong stochastic ordering between vectors and enables us to compare various increasing functionals of these vectors. Such constructions, called couplings, are possible under some assumptions on finite dimensional conditional distributions or other conditional characteristics of the corresponding distributions, which are sufficient but not necessary for strong stochastic ordering. For a general theory of coupling see Lindvall (1992).

Standard construction

This construction was given by Arjas and Lehtonen (1978).

Let $\mathbf{X} = (X_1, \ldots)$ and $\mathbf{Y} = (Y_1, \ldots)$ be random vectors, possibly with infinite number of coordinates. Let

$$F_1(x) = P(X_1 \le x);$$

$$F_j(x \mid x_1, \ldots, x_{j-1}) = P(X_j \le x \mid X_1 = x_1, \ldots, X_{j-1} = x_{j-1}),$$

for $j = 2, \ldots,\ x, x_j \in R$ where regular versions of conditional probabilities are taken (see Section 2.3 for conditioning). For $\mathbf{Y}$ we define the corresponding collection of G_j's. For brevity we introduce the following notation $\mathbf{x}^{|j} = (x_1, \ldots, x_j)$.

Theorem A. *If* $F_1 <_{st} G_1$ *and* $F_j(. \mid \mathbf{x}^{|j-1}) <_{st} G_j(. \mid \mathbf{y}^{|j-1})$, *for all* $\mathbf{x}^{|j-1} \le \mathbf{y}^{|j-1}, j = 2, \ldots$ *then there exist versions* $\tilde{\mathbf{X}}, \tilde{\mathbf{Y}}$ *of* $\mathbf{X}, \mathbf{Y}$ *on a common probability space such that* $\tilde{\mathbf{X}} \le \tilde{\mathbf{Y}}$ *a.s.*

Before we start the proof we recall a technical lemma (see Sec. 2.3 for a proof).

Lemma B. *Suppose that* $\mathbf{X}, \mathbf{Y}$ *are independent random vectors then*

$$P(\phi(\mathbf{X}, \mathbf{Y}) \in B \mid \mathbf{Y} = \mathbf{y}) = P(\phi(\mathbf{X}, \mathbf{y}) \in B) \ a.s.$$

for all Borel sets B *and Borel real functions* ϕ.

Proof of the theorem. We know from Section 1.1 that there exist a probability space $(\Omega, \mathcal{F}, P)$ and a sequence $(U_1, \ldots)$ of independent random variables on this space with identical uniform distributions on [0,1]. Define $\tilde{\mathbf{X}}$ as follows. For $\omega \in \Omega$

$$\tilde{X}_1(\omega) = F_1^{-1}(U_1(\omega));$$

$$\tilde{X}_j(\omega) = F_j^{-1}(U_j(\omega) \mid \tilde{X}_1(\omega), \ldots, \tilde{X}_{j-1}(\omega)), j \ge 2.$$

We have to prove that $\tilde{X}_j(\omega)$ is a Borel function of $\omega \in [0,1]$. We show that $F_j^{-1}(u \mid \mathbf{x}^{|j-1})$ is a Borel function of the variable $\mathbf{z} = (x_1, \ldots, x_{j-1}, u)$. Let $a \in R$. It is enough to prove that the following set is measurable

$$\{\mathbf{z} : F_j^{-1}(u \mid \mathbf{x}^{|j-1}) < a\} = \{\mathbf{z} : \inf\{t : F_j(t \mid \mathbf{x}^{|j-1}) \ge u\} < a\} =$$

$$= \{\mathbf{z} : \exists t < a, F_j(t \mid \mathbf{x}^{|j-1}) \geq u\} = \{\mathbf{z} : \exists n \in N, F_j(a - 1/n \mid \mathbf{x}^{|j-1}) \geq u\} =$$
$$= \{\mathbf{z} : \exists n \in N, \phi_{n,a}(\mathbf{x}^{|j-1}) \geq u\},$$

for $\phi_{n,a}(\mathbf{x}^{|j-1}) = P(X_j \leq a - 1/n \mid \mathbf{X}^{|j-1} = \mathbf{x}^{|j-1})$, which is a measurable function of $\mathbf{x}^{|j-1}$. Now

$$\{\mathbf{z} : \exists n \in N, \phi_{n.a}(\mathbf{x}^{|j-1}) \geq u\} = \cup_{n \in N}\{\mathbf{z} : \phi_{n,a}(\mathbf{x}^{|j-1}) - u \geq 0\}.$$

This set is measurable as a countable sum of measurable sets, which proves that $\tilde{X}_j(\omega)$ is a Borel function of ω. We prove that $\tilde{\mathbf{X}} =^d \mathbf{X}$. Of course $\tilde{X}_1 =^d X_1$. From Lemma B. we have

$$P(\tilde{X}_j \leq x \mid \tilde{X}_1 = x_1, \ldots, \tilde{X}_{j-1} = x_{j-1}) = P(F_j^{-1}(U_j \mid \mathbf{x}^{|j-1}) \leq x) =$$
$$= P(U_j \leq F_j(x \mid \mathbf{x}^{|j-1})) = P(X_j \leq x \mid X_1 = x_1, \ldots, X_{j-1} = x_{j-1}) \text{ a.s.}$$

From this we easily obtain $\tilde{\mathbf{X}} =^d \mathbf{X}$. Indeed, using induction assume that $\tilde{\mathbf{X}}^{|j-1} =^d \mathbf{X}^{|j-1}$. We have

$$P(\tilde{\mathbf{X}}^{|j} \leq \mathbf{x}^{|j}) = \int_{\mathbf{t}^{|j-1} \leq \mathbf{x}^{|j-1}} P(\tilde{X} \leq x_j \mid \tilde{\mathbf{X}}^{|j-1} = \mathbf{t}^{|j-1}) P^{\tilde{\mathbf{X}}^{|j-1}}(d\mathbf{t}^{|j-1}) =$$
$$= P(\mathbf{X}^{|j} \leq \mathbf{x}^{|j})$$

i.e. $\tilde{\mathbf{X}}^{|j} =^d n\mathbf{X}^{|j}$. From this we have $\tilde{\mathbf{X}} =^d \mathbf{X}$.
Using the same sequence $(U_1, \ldots)$ and the collection of $G_j^{\prime}s$ we can construct the corresponding vector $\tilde{\mathbf{Y}} =^d \mathbf{Y}$. We show that $\tilde{\mathbf{X}} \leq \tilde{\mathbf{Y}}$ a.s. Of course $\tilde{X}_1(\omega) \leq \tilde{Y}_1(\omega)$, $\omega \in \Omega$. Assume that $\tilde{\mathbf{X}}^{|j-1} \leq \tilde{\mathbf{Y}}^{|j-1}$ a.s. $j \geq 2$. We have

$$\tilde{X}_j(\omega) = F_j^{-1}(U_j(\omega) \mid \tilde{\mathbf{X}}^{|j-1}) \leq G_j^{-1}(U_j(\omega) \mid \tilde{\mathbf{Y}}^{|j-1}) = \tilde{Y}_j(\omega) \text{ a.s.}$$

From this we finally conclude that $\tilde{\mathbf{X}} \leq \tilde{\mathbf{Y}}$ a.s. □

Hazard rate construction

The next construction will be a generalization of one dimensional situation which we describe briefly below.

Let T be a nonnegative random variable possessing a density f with respect to the Lebesque measure. The failure rate of random life time T at moment t is given by

$$\lambda(t) = \lim_{\delta t \to 0+} P(t < T \leq t + \delta t)/\delta t P(T > t),$$

(we assume that the value of this expression is 0 if the denominator is equal to 0). The above limit, by the Lebesgue theorem on derivative of integrals, exists for almost all t and is equal to

$$\lambda(t) = f(t)/(1 - F(t))$$

where F denotes the cumulative distribution corresponding to f.
The function

$$\Lambda(t) = \int_0^t \lambda(s) ds$$

is called the **hazard function** of T.

Lemma C. *For* $t \geq 0$

$$\Lambda(t) = -\log(1 - F(t)).$$

Proof. This fact follows from the uniqueness of the solution of certain differential equation, it is possible however to use a more elementary argument. For t such that $F(t) < 1$, by integrating both sides of the above equation we get

$$\Lambda(t) = \int_0^t f(s)/(1 - F(s))ds.$$

The integrated function is equal a.s. with respect to the Lebesgue measure to the derivative of $-\log(1 - F(t))$. Therefore it is not possible here to apply the classical Newton-Leibniz formula. Consider the set of polynomials on $[0, t]$. It is a dense set in the space L_1 of all integrable functions on $[0, t]$. Let $w_n \to f$ in L_1. For

$$W_n(t) = \int_0^t w_n(s)ds$$

W_n converges to F uniformly on $[0, t]$. The function $1/(1 - x)$ is uniformly continuous on $[0, F(t)]$, hence $1/(1 - W_n)$ tends uniformly to $1/(1 - F)$ on $[0, t]$. From this fact we have that $\int_0^t w_n(s)/(1 - W_n(s))ds$ converges to $\int_0^t f(s)/(1 - F(s))$ ds. For the first of these integrals we can apply the fundamental formula of calculus, which gives the required equality. The most general version of the formula from the last lemma is called the Doleans-Dade (exponential) formula and can be proved by Fubini's theorem (see Jacod (1975) or Bremaud (1981)). □

The basic facts useful in constructions and having their multivariate analogs are as follows (see Theorem 1.2.F.).

Lemma D. *The random variable* $\Lambda(T)$ *is exponentially distributed with the mean 1.*

Lemma E. *If* E *is a random variable exponentially distributed with the mean 1 then*

$$T =^d \Lambda^{-1}(E)$$

Now we present a multivariate analogy of the above considerations. The idea of inverting the transformation $T \to E$ stems from Norros (1986) and was used in Shaked and Shanthikumar (1987). However, most of the following theorems are more explicite and use elementary methods.

We shall use the following notation. $\mathbf{T} = (T_1, \ldots, T_n)$ is a vector of nonnegative random variables which has a density $f(t_1, \ldots, t_n)$ with respect to the Lebesgue measure. For $J = \{j_1, \ldots, j_k\} \subseteq \{1, \ldots, n\}$ let $J^c = \{1, \ldots, n\} - J$, $\mathbf{t}_J = (t_{j_1}, \ldots, t_{j_k})$, $\max \mathbf{t}_J = \max\{t_{j_1}, \ldots, t_{j_k}\}$. The vector $\mathbf{e} = (1, \ldots, 1)$ has a dimensionally which depends on the context. The integral $\int_{t\mathbf{e}}^{\infty} d\mathbf{u}_J$ denotes $\int_t^\infty \ldots \int_t^\infty du_1 \ldots du_k$. By $f(t_1, \ldots, t_n) \mid_{t_i = t}$ we understand the function f with a fixed i-th coordinate t. We assume also that all the expressions considered are 0 if their denominators are 0.

We start with an elementary lemma.

Lemma F. *For*

$$f(\mathbf{t}_{J^c} \mid \mathbf{t}_j) = \frac{f(t_1,\ldots,t_n)}{\int_{-\infty}^{\infty} f(t_1,\ldots,t_n) d\mathbf{t}_{J^c}}$$

the function $\int_A f(\mathbf{t}_{J^c} \mid \mathbf{t}_J) d\mathbf{t}_{J^c}$ *is a version of the conditional probability* $P(T_{J^c} \in A \mid \mathbf{T}_J = \mathbf{t}_J)$

The version from the above lemma may not be a regular one, however for all $\mathbf{t}'_J s$ it is a probability measure.

The **conditional failure rate** of T_i, conditional on $\mathbf{T}_J = \mathbf{t}_J$, $J \subseteq \{1,\ldots,n\}, i \in J^c$ is $\lambda_i(t \mid \mathbf{T}_J = \mathbf{t}_J, \mathbf{T}_{J^c} > t\mathbf{e}) =$

$$= \lim_{\delta t \to 0+} \frac{1}{\delta t} \mathbf{Pr}(t < T_i \leq t + \delta t, \mathbf{T}_{J^c} > t\mathbf{e} \mid \mathbf{T}_J = \mathbf{t}_J) / \mathbf{Pr}(\mathbf{T}_{J^c} > t\mathbf{e} \mid \mathbf{T}_J = \mathbf{t}_J). \quad (1)$$

defined for $t \geq \max \mathbf{t}_J$ (for $J = \emptyset$ the condition $\mathbf{T}_J = \mathbf{t}_J$ is omitted).
The above definition would be incomplete without mentioning which version of the conditional probabilities is taken. A natural choice would be to take regular versions, however it will be easier to proceed with the version described in Lemma F..
We abbreviate $\lambda_i(t \mid \mathbf{T}_J = \mathbf{t}_J, \mathbf{T}_{J^c} > t\mathbf{e})$ to $\lambda_i(t \mid \mathbf{T}_J = \mathbf{t}_J, .)$ if no confusion is feared.

Lemma G. *The limit in (1) exists for almost all* $(t, \mathbf{t}_J)$ *and is a measurable function of* $(t, \mathbf{t}_J)$*, with*

$$\lambda_i(t \mid \mathbf{T}_J = \mathbf{t}_J, \cdot) = \frac{\int_{\mathbf{t}}^{\infty} f(t_1,\ldots,t_n) \mid_{t_i = t} d\mathbf{t}_{J^c \setminus \{i\}}}{\int_{\mathbf{t}}^{\infty} f(t_1,\ldots,t_n) d\mathbf{t}_{J^c}}, \quad a.s.$$

Proof. The expression under the lim in (1) equals

$$\frac{1}{\delta t} \frac{\int_t^{t+\delta t} \int_{\mathbf{t}}^{\infty} f(\mathbf{t}_{J^c} \mid \mathbf{t}_J) d\mathbf{t}_{J^c \setminus \{i\}} dt_i}{\int_{\mathbf{t}}^{\infty} f(\mathbf{t}_{J^c} \mid \mathbf{t}_J) d\mathbf{t}_{J^c}},$$

which can be written as

$$\frac{1}{\delta t} \frac{\int_t^{t+\delta t} \int_{\mathbf{t}}^{\infty} f(t_1,\ldots,t_n) d\mathbf{t}_{J^c \setminus \{i\}} dt_i}{\int_{\mathbf{t}}^{\infty} f(t_1,\ldots,t_n) d\mathbf{t}_{J^c}}. \quad (2)$$

Now we substitute $t_i = s_i$, $t_j = s_j$ for $j \in J^c \setminus \{i\}$. The resulting function after this exchange we denote by $g(\mathbf{s}_{J^c}, \mathbf{t}_J)$. Jacobian of the transformation $(s_i, s_{j_1}, \ldots, s_{j_k}) \to (s_i, s_i + s_{j_1}, \ldots, s_i + s_{j_k})$, where $J^c = \{i, j_1, \ldots, j_k\}$ equals 1, therefore the numerator in (2) is equal to

$$I(t, \delta t) = \frac{1}{\delta t} \int_t^{t+\delta t} \int_{(t-s_i)\mathbf{e}}^{\infty} g(\mathbf{s}_{J^c}, \mathbf{t}_J) d\mathbf{s}_{J^c \setminus \{i\}} ds_i.$$

Let

$$I_0(t, \delta t) = \frac{1}{\delta t} \int_t^{t+\delta t} \int_{\mathbf{0}}^{n\infty} g(\mathbf{s}_{J^c}, \mathbf{t}_J) d\mathbf{s}_{J^c \setminus \{i\}} ds_i.$$

and

$$I_k(t, \delta t) = \frac{1}{\delta t} \int_t^{t+\delta t} \int_{(-1/k)\mathbf{e}}^{\infty} g(\mathbf{s}_{J^c}, \mathbf{t}_J) d\mathbf{s}_{J^c \setminus \{i\}} ds_i.$$

for $k \geq 1$. In the next lemma we prove that the limits with $\delta t \to 0$ of I_0 and I_k do exist for almost all $(t, \mathbf{t}_J)$ and are equal

$$\int_0^\infty [g(\mathbf{s}_{J^c}, \mathbf{t}_J) \mid_{s_i=t}] d\mathbf{s}_{J^c \setminus \{i\}},$$

$$\int_{(-1/k)\mathbf{e}}^\infty [g(\mathbf{s}_{J^c}, \mathbf{t}_J) \mid_{s_i=t}] d\mathbf{s}_{J^c \setminus \{i\}},$$

respectively, and are measurable.
Now, since we have

$$I_0(t, \delta t_m) \leq I(t, \delta t_m) \leq I_k(t, \delta t_m),$$

for a sequence of $\delta t_m \to 0+$, from the Lebesgue bounded convergence theorem we obtain

$$\lim_{\delta t \to 0+} I(t, \delta t) = \int_0^\infty [g(\mathbf{s}_{J^c}, \mathbf{t}_J) \mid_{s_i=t}] d\mathbf{s}_{J^c \setminus \{i\}}.$$

which completes the proof. □

Lemma H. *Suppose that $g : R \times R^s \to R$ is a nonnegative integrable function. For $h : R \times R^s \to R$ given by*

$$h(t, \mathbf{y}) = \lim_{\delta t \to 0+} (1/\delta t) \int_t^{t+\delta t} g(x, \mathbf{y}) dx$$

we have $h(t, \mathbf{y}) = g(t, \mathbf{y})$ a.s., moreover h is measurable

Proof. Let

$$H(I, y) = \frac{1}{\mid I \mid} \int_I g(x, y) dx,$$

where I is a closed interval in R and $\mid I \mid$ its Lebesgue measure. From the Fubini theorem $g(\cdot, y)$ is integrable for almost all y. Without loss of generality we assume that it is integrable for all y, then for each fixed y the limit $\lim_{I \to \{t\}} H(I, y)$ exist and equals $g(t, y)$, for almost all t, from the theorem on differentiating of Lebesgue integrals. The set of (t, y) for which the limit exist can be proved to be measurable, but we omit this measurability problem. □

From this lemma we can see that the failure rates of more than one components at the same time are 0. More precisely.

Lemma I. *If $i, j \in J^c, i \neq j$ then*

$$\lim_{\delta t \to 0+} (1/\delta t) P(t < T_i \leq t + \delta t, t < T_j \leq t + \delta t, \mathbf{T}_{J^c} > t\mathbf{e} \mid \mathbf{T}_J = \mathbf{t}_J) = 0$$

for almost all $(t, \mathbf{t}_J)$.

Proof. The proof is similar to that of Lemma G. and we omit it.

The **conditional hazard function** of T_i on $[\max \mathbf{t}_J, t + \max \mathbf{t}_J]$ conditioned on $\mathbf{T}_J = \mathbf{t}_J$ is

$$\Lambda_i(t \mid \mathbf{T}_J = \mathbf{t}_J) = \int_{\max \mathbf{t}_J}^{t + \max \mathbf{t}_J} \lambda_i(u \mid \mathbf{T}_J = \mathbf{t}_J, \cdot) du$$

Lemma G. and the Fubini theorem imply the existence of Λ and its measurability.
Let

$$F_i(t_i \mid \mathbf{t}_{\{1,\ldots,i-1\}}) = \frac{\mathbf{Pr}(\mathbf{T}_{\{i,\ldots,n\}} > t_i\mathbf{e} \mid \mathbf{T}_{\{1,\ldots,i-1\}} = \mathbf{t}_{\{1,\ldots,i-1\}})}{\mathbf{Pr}(\mathbf{T}_{\{i,\ldots,n\}} > t_{i-1}\mathbf{e} \mid \mathbf{T}_{\{1,\ldots,i-1\}} = \mathbf{t}_{\{1,\ldots,i-1\}}},$$

for $i = 1, \ldots, n$, and $0 = t_0 < t_1 < \ldots < t_i$.
From Lemma F. it is clear that

$$F_i(t_i \mid \mathbf{t}_{\{1,\ldots,i-1\}}) = \frac{\int_{t_i\mathbf{e}}^{\infty} f(t_1, \ldots, t_n) d\mathbf{t}_{\{i,\ldots,n\}}}{\int_{t_{(i-1)}\mathbf{e}}^{\infty} f(t_1, \ldots, t_n) d\mathbf{t}_{\{i,\ldots,n\}}} \tag{3}$$

We shall need the following formula.

Lemma J. *For almost all* $(t_1, \ldots, t_n)$ *such that* $0 = t_0 < t_1 < \ldots < t_n$
$f(t_1, \ldots, t_n) =$

$$= \prod_{i=1}^{n} (\lambda_i(t_i \mid \mathbf{T}_{\{1,\ldots,i-1\}} = \mathbf{t}_{\{1,\ldots,i-1\}}, \cdot) \times \exp(-\sum_{k=i}^{n} \Lambda_k(t_i - t_{i-1} \mid \mathbf{T}_{\{1,\ldots,i-1\}} = \mathbf{t}_{\{1,\ldots,i-1\}})))$$

Proof. We prove first

$$f(t_1, \ldots, t_n) = \prod_{i=1}^{n} \lambda_i(t_i \mid \mathbf{T}_{\{1,\ldots,i-1\}} = \mathbf{t}_{\{1,\ldots,i-1\}}, \cdot) F_i(t_i \mid \mathbf{t}_{\{1,\ldots,i-1\}}). \tag{4}$$

If we apply Lemma G. and formula (3) then we get (4), provided the integrals involved are positive. On the set where one of these integrals is zero f is also zero. Indeed, for a fixed i the considered integrals are of the form

$$I(t_1, \ldots, t_n) = \int_{t_i\mathbf{e}}^{\infty} f(t_1, \ldots, t_i, u_{i+1}, \ldots, u_n) d\mathbf{u}_{\{i+1,\ldots,n\}},$$

or

$$J(t_1, \ldots, t_n) = \int_{t_i\mathbf{e}}^{\infty} f(t_1, \ldots, t_{i-1}, u_i, \ldots, u_n) d\mathbf{u}_{\{i,\ldots,n\}}.$$

Let $A = \{(t_1, \ldots, t_n) : I(t_1, \ldots, t_n) = 0\}$. Now

$$\int_0^\infty \int_{t_1}^\infty \cdots \int_{t_{n-1}}^\infty I_A(t_1, \ldots, t_n) f(t_1, \ldots, t_n) dt_n \cdots dt_2 dt_1 \leq$$

$$\int_0^\infty \int_{t_1}^\infty \cdots \int_{t_{i-1}}^\infty [I_A(t_1, \ldots, t_n) \int_{t_i\mathbf{e}}^{\infty} f(t_1, \ldots, t_i, u_{i+1}, \ldots, u_n) d\mathbf{u}_{\{i+1,\ldots,n\}}] dt_i \cdots dt_2 dt_1 = 0.$$

Consequently f is a.s. zero on A.
We show now that

$$F_i(t_i \mid \mathbf{t}_{\{1,\ldots,i-1\}}) = \exp[-\sum_{k=i}^{n} \Lambda_k(t_i - t_{i-1} \mid \mathbf{T}_{\{1,\ldots,i-1\}} = \mathbf{t}_{\{1,\ldots,i-1\}})].$$

If the denominator in (3) is positive, F_i has, as a function of t_i, $t_i \geq t_{i-1}$, all properties of tail distributions possessing a density with respect to Lebesgue measure. For $t \leq t_{i-1}$ its failure rate is zero. For $t > t_{i-1}$ its failure rate equals $\sum_{k=i}^{n} \lambda_k(t \mid \mathbf{T}_{\{1,\ldots,i-1\}} = \mathbf{t}_{\{1,\ldots,i-1\}}, \cdot)$. Indeed, consider

$$\lim_{\delta t \to 0+} \frac{1}{\delta t}[F_i(t \mid \mathbf{t}_{\{1,\ldots,i-1\}}) - F_i(t+\delta t \mid \mathbf{t}_{\{1,\ldots,i-1\}})].$$

The event $\{\mathbf{T}_{\{i,\ldots,n\}} > t\mathbf{e} \setminus \mathbf{T}_{\{i,\ldots,n\}} > (t+\delta t)\mathbf{e}\}$ we represent as

$$\bigcup_{\emptyset \neq J \subset \{i,\ldots,n\}} \{t\mathbf{e} < \mathbf{T}_J \leq (t+\delta t)\mathbf{e},\ \mathbf{T}_{\{i,\ldots,n\}\setminus J} > (t+\delta t)\mathbf{e}\}.$$

Events with more than one element in J lead to zero limits. The event $\{t < \mathbf{T}_k \leq t+\delta t,\ \mathbf{T}_{\{i,\ldots,n\}\setminus\{k\}} > (t+\delta t)\mathbf{e}\}$ can be written as

$$\{t < \mathbf{T}_k \leq t+\delta t,\ \mathbf{T}_{\{i,\ldots,n\}\setminus\{k\}} > t\mathbf{e}\} \setminus \bigcup_J \{t\mathbf{e} < \mathbf{T}_J \leq (t+\delta t)\mathbf{e},\ \mathbf{T}_{\{i,\ldots,n\}\setminus J} > (t+\delta t)\mathbf{e}\},$$

where the sum is over at least two element sets. The first event gives $\lambda_k(t \mid \mathbf{T}_{\{1,\ldots,i-1\}} = \mathbf{t}_{\{1,\ldots,i-1\}}, \cdot)$ the rest is zero. □

Note that similar equalities are valid for other permutations of $t_1, \ldots, t_n$.

The **conditional total hazard function** of T_i on $[0,t]$, conditioned on $\mathbf{T}_{j_1} = t_{j_1}, \ldots, T_{j_k} = t_{j_k}, 0 = t_{j_0} < t_{j_1} < \ldots < t_{j_k} < t$, is

$\Psi_{i|j_1,\ldots,j_{k-1}}(t \mid t_{j_1}, \ldots, t_{j_{k-1}}) =$

$$= \sum_{m=1}^{k-1} \Lambda_i(t_{j_m} - t_{j_{m-1}} \mid \mathbf{T}_{j_1} = t_{j_1}, \ldots, T_{j_{m-1}} = t_{j_{m-1}}) + \Lambda_i(t - t_{j_{k-1}} \mid \mathbf{T}_{j_1} = t_{j_1}, \ldots, T_{j_{k-1}} = t_{j_{k-1}})$$

Define in addition

$$A_i(t \mid t_{j_1}, \ldots, t_{j_n}) = \sum_{k=1}^{n} I_{(t_{j_{k-1}}, t_{j_k}]}(t) \Psi_{i|j_1,\ldots,j_{k-1}}(t \mid t_{j_1}, \ldots, t_{j_{k-1}})$$

for $0 = t_{j_0} < t_{j_1} < \ldots < t_{j_n}$.
The basic theorem for our construction is as follows.

Theorem K. *The random variables given by*

$$E_i(\omega) = A_i(T_i(\omega) \mid T_{j_1}(\omega), \ldots, T_{j_n}(\omega)),$$

where $j_1, \ldots, j_n$ are random indices which indicate the increasing rearrangement of $T_1(\omega), \ldots, T_n(\omega)$, are independent and exponentially distributed with mean 1.

Proof. We derive the joint characteristic function of $(E_1, \ldots, E_n)$. Proceeding by induction, we see that for $n = 1$ the theorem follows from Lemma D. . Now, in
$E(\exp(\sum_{j=1}^{n} iu_j E_j)) = \int_0^\infty \exp(\sum_{j=1}^{n} iu_j A_j(t_j \mid t_{j_1}, \ldots, t_{j_n})) f(t_1, \ldots, t_n) d\mathbf{t}_{\{1,\ldots,n\}}$
we split the area of integration into the sets

$$D_j = \{(t_1, \ldots, t_n) : t_j = \min\{t_1, \ldots, t_n\}\},\ j = 1, \ldots, n.$$

We calculate the case $j = 1$.

For a fixed t_1 we define a measure P_{t_1} by putting
$P_{t_1}(T_2 \le t_2, \ldots, T_n \le t_n) =$

$$= \int_{t_1}^{t_2} \ldots \int_{t_1}^{t_n} f(t_1, \ldots, t_n) dt_n \ldots dt_2 / \int_{t_1\mathbf{e}}^{\infty} f(t_1, \ldots, t_n) d\mathbf{t}_{\{2,\ldots,n\}}.$$

Assume that the denominator of the above expression is positive. Then this formula defines a probability measure concentrated on $[t_1, \infty)^{n-1}$, with a density with respect to the Lebesgue measure

$$g_{t_1}(t_2, \ldots, t_n) = f(t_1, \ldots, t_n) / \int_{t_1\mathbf{e}}^{\infty} f(t_1, \ldots, t_n) d\mathbf{t}_{\{2,\ldots,n\}}$$

for $t_1 \le \min\{t_2, \ldots, t_n\}$.
If $J \subseteq \{2, \ldots, n\}, i \in J^c, i > 1$, then from the above equality we have

$$\lambda_{t_1,i}(t \mid T_J = \mathbf{t}_J, \mathbf{T}_{\{2,\ldots,n\}-J} > t\mathbf{e}) = \lambda_i(t \mid T_J = \mathbf{t}_J, \mathbf{T}_{\{2,\ldots,n\}-J} > t\mathbf{e})$$

for almost all $(t, \mathbf{t}_J)$ where functions with the index t_1 correspond to the distribution P_{t_1}.
From Lemma J. we have now

$$f(t_1, \ldots, t_n) = \lambda_1(t_1) \exp(-\sum_{j=1}^{n} \Lambda_j(t_1)) g_{t_1}(t_2, \ldots, t_n).$$

At the same time
$A_1(t_1 \mid t_{j_1}, \ldots, t_{j_n}) = \Lambda_1(t_1)$,
$A_j(t_j \mid t_{j_1}, \ldots, t_{j_n}) = \Lambda_1(t_1) + A_{t_1,j}(t_j \mid t_{j_1}, \ldots, t_{j_n}), j = 2, \ldots, n.$
Thus the integral over D_1 is equal

$$\int_0^{\infty} \lambda_1(t_1) \exp(\sum_{j=1}^{n} (iu_j - 1) \Lambda_j(t_1)) \phi_{t_1}(u_2, \ldots, u_n) dt_1$$

where

$$\phi_{t_1}(u_2, \ldots, u_n) =$$

$$= \int_{t_1\mathbf{e}}^{\infty} \exp(\sum_{j=2}^{n} iu_j A_{t_1,j}(t_j \mid t_{j_1}, \ldots, t_{j_n})) g_{t_1}(t_2, \ldots, t_n) d\mathbf{t}_{\{2,\ldots,n\}}.$$

From the inductive assumption

$$\phi_{t_1}(u_2, \ldots, u_n) = \prod_{j=2}^{n} 1/(1 - iu_j)$$

i.e. ϕ_{t_1} is the characteristic function of the vector of independent exponentially distributed random variables. Hence our integral takes on the following form

$$\prod_{j=1}^{n} 1/(1 - iu_j) \int_0^{\infty} (1 - iu_1) \lambda_1(t_1) \exp(\sum_{j=1}^{n} (iu_j - 1) \Lambda_j(t_1)) dt_1.$$

After adding all the areas of integration we get

$$E(\exp(\sum_{j=1}^{n} iu_jE_j)) = \prod_{j=1}^{n} 1/(1-iu_j) \int_0^{\infty} (\sum_{j=1}^{n}(1-iu_j)\lambda_j(t)) \exp(\sum_{j=1}^{n}(iu_j-1)\Lambda_j(t))dt$$

which can be rewritten as

$$\int_0^{\infty} \exp(-\sum_{j=1}^{n}\Lambda_j(t))(\cos(\sum_{j=1}^{n} u_j\Lambda_j(t))(\sum_{j=1}^{n}\lambda_j(t)) + \sin(\sum_{j=1}^{n} u_j\Lambda_j(t))(\sum_{j=1}^{n} u_j\lambda_j(t)))dt$$

$$-i\int_0^{\infty} \exp(-\sum_{j=1}^{n}\Lambda_j(t))(\cos(\sum_{j=1}^{n} u_j\Lambda_j(t))(\sum_{j=1}^{n} u_j\lambda_j(t)) - \sin(\sum_{j=1}^{n} u_j\Lambda_j(t))(\sum_{j=1}^{n}\lambda_j(t)))dt.$$

Now similarly as in Lemma D. we can show that the first integral above equals 1, the second one 0. □

The mapping $(T_1,\ldots,T_n) \to (E_1,\ldots,E_n)$ can be in a sense inverted. Let $j_1,\ldots,j_n$ be such that $T_{j_1} < \ldots < T_{j_n}$. Then we have

$E_{j_1} = \Psi_{j_1}(T_{j_1})$,

$E_{j_2} = \Psi_{j_2|j_1}(T_{j_2} \mid T_{j_1})$,

$\vdots$

$E_{j_n} = \Psi_{j_n|j_1,\ldots,j_n}(T_{j_n} \mid T_{j_1},\ldots,T_{j_{n-1}})$

or alternatively

$E_{j_1} = \Lambda_{j_1}(T_{j_1})$,

$E_{j_2} = \Lambda_{j_2}(T_{j_1}) + \Lambda_{j_2}(T_{j_2} - T_{j_1} \mid T_{j_1})$,

$\vdots$

$E_{j_n} = \Lambda_{j_n}(T_{j_1}) + \ldots + \Lambda_{j_n}(T_{j_n} - T_{j_{n-1}} \mid T_{j_1},\ldots,T_{j_{n-1}})$.

We have of course

$$\partial/\partial t_i \Lambda_{j_i}(t_i - t_{i-1}) \mid \mathbf{T}_{\{j_1,\ldots,j_{i-1}\}} = \mathbf{t}_{\{1,\ldots,i-1\}}) =$$

$$= \lambda_{j_i}(t_i \mid \mathbf{T}_{\{j_1,\ldots,j_{i-1}\}} = \mathbf{t}_{\{1,\ldots,i-1\}}, \cdot)$$

for t_i a.s. and $t_1 < \ldots < t_i$. If this derivative is positive then t_i is a point of increase for

$$\Lambda_{j_i}(t_i - t_{i-1}) \mid \mathbf{T}_{\{j_1,\ldots,j_{i-1}\}} = \mathbf{t}_{\{1,\ldots,i-1\}})$$

If we substitute $t_k = T_{j_k}(\omega)$, $k = 1,\ldots,i$ then this derivative is positive with probability 1. For all ω in a set of measure 1 thus

$T_{j_1}(\omega) = \min\{\Psi_j^{-1}(E_j(\omega) : j \in \{1,\ldots,n\}\}$,

$T_{j_2}(\omega) = \min\{\Psi_{j|j_1}^{-1}(E_j(\omega) \mid T_{j_1}(\omega)) : j \in \{1,\ldots,n\} - \{j_1\}\}$,

$\vdots$

$T_{j_n}(\omega) = \Psi_{j_n|j_1,\ldots,j_{n-1}}^{-1}(E_{j_n}(\omega) \mid T_{j_1}(\omega),\ldots,T_{j_{n-1}}(\omega))$.

Here all the minima are attained for just one index (k-th for j_k).

This situation allows us to use the following construction.
Let $\mathbf{E} = (E_1, \ldots, E_n)$ be a vector of independent random variables exponentially distributed with means 1. We define a vector $\mathbf{T}' = (T_1', \ldots, T_n')$ in the consecutive n steps.
Step 1. Let

$$t_1 = \min\{\Psi_j^{-1}(E_j(\omega)) : j \in \{1, \ldots, n\}\}$$

and put $T_{j_1}'(\omega) = t_1$;
Step k. If $T_{j_1}'(\omega) = t_1, \ldots, T_{j_{k-1}}'(\omega) = t_{k-1}$, let

$$t_k = \min\{\Psi_{j|j_1,\ldots,j_{k-1}}^{-1}(E_j(\omega) \mid t_1, \ldots, t_{k-1}) : j \in \{j_1, \ldots, j_{k-1}\}^c\}$$

and put $T_{j_k}'(\omega) = t_k$;
Step n. For the last index j_n we put

$$T_{j_n}'(\omega) = \Psi_{j_n|j_1,\ldots,j_{n-1}}^{-1}(E_{j_n}(\omega) \mid t_1, \ldots, t_{n-1}).$$

From the above construction we have

Theorem L. *Under the above conditions*

$$\mathbf{T} =_{st} \mathbf{T}'$$

2.3 Conditioning

In an elementary probability theory the conditional probability with respect to a set with a positive measure plays an important role. To be more specific, consider a probability space $(\Omega, \mathcal{F}, P)$, and assume that for $A \in \mathcal{F}, P(A) > 0$. The conditional probability with respect to A is a measure defined by $P(B \mid A) = P(A \cap B)/P(A)$, for $B \in \mathcal{F}$. A generalization of this concept to conditioning under a collection of sets proved to be fundamental in introducing several important classes of stochastic processes.

We begin by describing what we understand under the term "stochastic process". Consider a family of random variables $\{X_t, t \in T\}$, on the same probability space $(\Omega, \mathcal{F}, P)$, where T is an arbitrary index set. We usually assume that $T = R$ or $T = N$. Such a family we call **a random function.** For $\omega \in \Omega$, $X_t(\omega)$ is a function of $t \in T$ called **a trajectory** of $\{X_t, t \in T\}$. Consider another family $\{\tilde{X}_t, t \in T\}$ on perhaps a different probability space $(\tilde{\Omega}, \tilde{\mathcal{F}}, \tilde{P})$ such that

$$\tilde{P}(\tilde{X}_{t_1} \in B_1, \ldots, \tilde{X}_{t_k} \in B_k) = P(X_{t_1} \in B_1, \ldots, X_{t_k} \in B_k),$$

for all $t_1 < \ldots < t_k, k \geq 1, B_1, \ldots, B_k \in \mathcal{B}^1$. We say than that the random functions $\{X_t\}$ and $\{\tilde{X}_t\}$ are stochastically equivalent and we write $\{X_t\} =^d \{\tilde{X}_t\}$. $\{\tilde{X}\}$ is called **a version** of $\{X_t\}$ and vice versa. The relation $=^d$ is the equivalence type relation. By a stochastic process $(X_t, t \in T)$ we understand the corresponding to a random function $\{X_t\}$ equivalence class with respect to the $=^d$ relation. With this definition we can view a stochastic process as a family of finite dimensional distributions $\{P(X_{t_1} \in B_1, \ldots, X_{t_k} \in B_k)\}$. The fact that one stochastic process can have versions with different trajectory properties will be clear from the following example.

Example. For $(\Omega, \mathcal{F}, P) = ([0,1], \mathcal{B}^1, \ell), T = [0,1]$, define $X_t(\omega) = 0$ for all $\omega, t \in [0,1]$, and $\tilde{X}_t(\omega) = 1$ if $t = \omega$, zero otherwise. Of course $\{X_t\} =^d \{\tilde{X}_t\}$ but the trajectories of X_t are continuous while the trajectories of $\tilde{X}$ have discontinuities at $t = \omega$.

Martingale theory, based on conditioning, is an essential tool in the analysis of Markov processes and point processes. Therefore, before continuing with Markov processes and point processes we give in this section a brief account of conditioning with respect to σ-fields, trying to illustrate this concept rather than develop a general theory. For a more systematic approach see e.g. Billingsley (1986). We begin by describing this concept for a finite disjoint collection of sets $A_1, \ldots, A_n \in \mathcal{F}$, such that $P(A_i) > 0, i = 1, \ldots, n$. and $A_1 \cup \ldots \cup A_n = \Omega$. Denote by $\mathcal{G}$ the smallest σ-field containing $A_1, \ldots, A_n$. Now for each $B \in \mathcal{F}$ we define a simple random variable which assumes the value $P(B \mid A_i)$ on the set $A_i, i = 1, \ldots$,n. Because each A_i has a positive probability the values $P(B \mid A_i)$ are well defined. This random variable is usually denoted by $P(B \mid \mathcal{G})$. Using indicator functions we give a compact formula for it

$$P(B \mid \mathcal{G})(\omega) = \sum_{i=1}^{n} P(B \mid A_i) I_{A_i}(\omega),$$

where I_{A_i} denotes the indicator function of A_i. The random variable $P(B \mid \mathcal{G})$ is called the conditional probability of B given the σ-field $\mathcal{G}$. Such σ-fields have a very simple structure, they consist of all the unions of the A_i's. A further step in generalization of conditioning is to take an infinite partition of Ω, $A_1, \ldots$, not assuming that for all i, $P(B \mid A_i) > 0$. The only modification in the above definition is that we allow the random variable $P(B \mid \mathcal{G})$ to take arbitrary constant values on sets A_i, for which $P(A_i) = 0$. On other sets it is defined by the above compact formula. In this situation we have freedom of choosing arbitrary values on sets with measure zero. However, when writing $P(B \mid \mathcal{G})$, we think of the whole family of functions with different values on zero measure sets. Each particular random variable $P(B \mid \mathcal{G})(\omega)$ is then a version of the conditional probability $P(B \mid \mathcal{G})$. These versions can be different only on sets with measure zero. Note that the random variable $P(B \mid \mathcal{G})$ has a finite expectation

$$EP(B \mid \mathcal{G}) = P(B),$$

and moreover for each set $A \in \mathcal{G}$ we have

$$E(P(B \mid \mathcal{G}) I_A) = P(B \cap A). \tag{1}$$

The property (1) is what we could expect from a general conditional probability to fulfill, and resembles the basic concept of conditioning. It is not surprising that this property together with some other conditions can be used to define the most general conditioning concept for arbitrary σ-fields $\mathcal{G}$.

Definition A. *A random variable $P(B \mid \mathcal{G})$ is a version of the conditional probability of B given a σ-field $\mathcal{G}$ if it is $\mathcal{G}$-measurable, has a finite expectation and fulfills (1), for all $A \in \mathcal{G}$.*

There will in general be many such random variables, but any two of them are equal with probability 1. The Radon-Nikodym theorem can be applied to prove the existence of $P(B \mid \mathcal{G})$ (see e.g. Billingsley (1986)). However in many particular cases we can find such conditional probabilities explicitly.

Example. $(\Omega, \mathcal{F}, P) = (R^2, \mathcal{B}^2, P)$, where P has a density with respect to Lebesgue measure $f(x,y)$. Let $B = R \times F$, for $F \in \mathcal{B}^1$, and $\mathcal{G} = \sigma((a,b) \times R : a < b)$. We will check from Definition A. that for $\omega = (x,y) \in R^2$

$$P(B \mid \mathcal{G})(x,y) = \int_F f(x,t)dt / \int_R f(x,t)dt.$$

Because $P(B \mid \mathcal{G})$ does not depend on y, it is $\mathcal{G}$ measurable. We have to show the condition (1) i.e.

$$E(P(B \mid \mathcal{G})I_A) = P(B \cap A),$$

for $A \in \mathcal{G}$. Take arbitrary $A \in \mathcal{G}$ of the form $A = E \times \mathrm{R}$. The above equation can be rewritten in the following form

$$\int_E \int_R [\int_F f(x,t)dt / \int_R f(x,t)dt] f(x,y)dxdy = P(E \times F),$$

which is fulfilled from Fubini's theorem.

If the σ-field $\mathcal{G}$ is generated by a random variable i.e. $\mathcal{G} = \sigma(X)$ ($\sigma(X) = \sigma(X^{-1}(B)$, $B \in \mathcal{B}^1$), for a random variable X, than we write $P(B \mid X)$ rather than $P(B \mid \mathcal{G})$. In this case the following lemma sheds some additional light on conditioning.

Lemma B. *If Z is a random variable $\sigma(X)$ measurable then $Z = \phi(X)$, for some Borel function ϕ.*

Proof. Suppose Z is a simple random variable of the form

$$Z = \sum_i a_i I_{F_i}, \; F_i \in \sigma(X)$$

where $F_i = X^{-1}(A_i)$ for some $A_i \in \mathcal{B}^1$. Now we can write Z as

$$Z(\omega) = \sum_i a_i I_{A_i}(X(\omega)).$$

If we take $\phi(x) = \sum_i a_i I_{A_i}(x)$, which is measurable, then $Z = \phi(X)$. If Z is a general random variable then we apply a standard approximation argument. □

Because $P(B \mid X)$ is from definition $\sigma(X)$ measurable we have that $P(B \mid X) = \phi(X)$, for some Borel function ϕ. The function ϕ can be sometimes explicitly found, as in the above example, namely if we additionally assume in this example that the density f is a density of a distribution $P^{X,Y}$ of a pair of random variables (X, Y) on R^2 then

$$\phi(x) = \int_F f(x,t)dt / \int_R f(x,t)dt.$$

Usually we use $P(Y \in F \mid X = x)$ to denote such a function ϕ. It is clear that the above example can be generalized to give

$$P(Y_1 \in B_1, \ldots, Y_n \in B_n \mid X_1 = x_1, \ldots, X_n = x_n) =$$

$$\int_{B_1,\ldots,B_n} f(y_1, \ldots, y_n, x_1, \ldots, x_n) dy_1 \ldots dy_n / \int_{R^n} f(y_1, \ldots, y_n, x_1, \ldots, x_n) dy_1 \ldots dy_n,$$

where f is a joint density of a random vector $(\mathbf{X}, \mathbf{Y})$ of the dimension $2n$, and the left hand side of this equality denotes a function $\phi(\mathbf{x})$, for which we have

$$P(\mathbf{Y} \in B_1 \times \ldots \times B_n \mid \sigma(\mathbf{X})) = \phi(\mathbf{X}) \text{ a.s.}$$

Example. Let X, Y be independent with the same positive and continuous distribution function F.
We check that

$$P(X \le x \mid \max(X,Y))(\omega) =$$
$$I_{(\max(X,Y) \le x)}(\omega) + I_{(\max(X,Y) > x)}(\omega) F(x)/2F(\max(X(\omega), Y(\omega)).$$

Indeed, we have only to verify that

$$E((I_{(\max(X,Y) \le x)}(\omega) + I_{(\max(X,Y) > x)}(\omega) F(x)/2F(\max(X(\omega), Y(\omega)) I_G) =$$
$$= P((X \le x) \cap G),$$

for $G \in \sigma(\max(X,Y))$, because the measurability condition is trivially fulfilled. A standard way of checking such a condition for arbitrary G is to assume first that G has a special simple form, and than use a π-class argument (see e.g. Billingsley (1986)). In the case of real valued random variables we take usually $G = (\max(X,Y) \le m), m \in R$, i.e. the inverses by the given random variable of intervals of the form $(-\infty, m]$. If $x < m$ we have with $M = \max(X,Y)$

$$E(I_{(M \le x)} I_{(M \le m)} + I_{(x < M \le m)} F(x)/2F(M)) =$$
$$P(M \le x) + (1/2)F(x) E(I_{(x<M\le m)} 1/F(M)) =$$
$$= P(M \le x) + (1/2)F(x) \int\int_{x < \max(u,v) \le m} 1/F(\max(u,v)) dF(u) dF(v) =$$
$$= F^2(x) + (1/2)F(x) 2[F(m) - F(x)] = F(m)F(x) = P(M \le m, X \le x)$$

which is obtained after applying the Fubini theorem, independence of X and Y, and the continuity property of F. The case $x \ge m$ is similar.

Example. If X, Y are independent than

$$P(\max(X,Y) \le m \mid X)(\omega) = \delta_{X(\omega)}(m) P(Y \le m) \text{ a.s.}$$

This is a consequence of the fact that for arbitrary measurable function $\psi(x,y)$, and independent X, Y, we have for $B \in \mathcal{B}^2$

$$P(\psi(X,Y) \in B \mid X = x) = P(\psi(x,Y) \in B) \text{ a.s.}$$

Indeed

$$E(P(\psi(X,Y) \in B \mid X) I_{(X \in A)}) =$$

$$\int_A P(\psi(x,Y) \in B) dF_X(x) = \int_A \int_R I_{(y:\psi(x,y)\in B)}(y) dF_Y(y) dF_X(x) =$$

$$= \int\int_{R\times R} I_{((x,y):\psi(x,y)\in B, x\in A)}(x,y) dF_{X,Y}(x,y) = P(\psi(X,Y) \in B, X \in A),$$

hence (1) is fulfilled. Now $\psi(x,y) = \max(x,y)$, and

$$P(\max(x,Y) \leq m) = \delta_x(m) P(Y \leq m),$$

where δ_x denotes the atom distribution at x.

In an analogous way to conditional probabilities we define conditional expected values.

Definition C. *A random variable $E(X \mid \mathcal{G})$ is a version of the expected value of a random variable X given a $\sigma-$ field $\mathcal{G}$ if it is $\mathcal{G}$ measurable, has a finite expectation and*

$$E(E(X \mid \mathcal{G}) I_A) = E(X I_A),$$

for all $A \in \mathcal{G}$.

If $X = I_B$, for $B \in \mathcal{F}$ then $E(I_B \mid \mathcal{G}) = P(B \mid \mathcal{G})$ a.s.

Conditional expectations have all nice properties of usual expectations, and some properties which allow us treat them as projections.

Theorem D. *Assume that X_n, X, Y are random variables with finite expectations, and $\mathcal{F}$, $\mathcal{G}$ are sigma fields. Then*

(i) $E(X + Y \mid \mathcal{F}) = E(X \mid \mathcal{F}) + E(Y \mid \mathcal{F})$;

(ii) *If $X_n \to X$, a.s. in the increasing way then $E(X_n \mid \mathcal{F}) \to E(X \mid \mathcal{F})$ a.s. , $n \to \infty$;*

(iii) $E(E(X \mid \mathcal{F}) \mid \mathcal{F}) = E(X \mid \mathcal{F})$ *a.s.;*

(iv) $E(E(X \mid \mathcal{F}) \mid \mathcal{G}) = E(E(X \mid \mathcal{G}) \mid \mathcal{F}) = E(X \mid \mathcal{F})$ *a.s.,provided $\mathcal{F} \subseteq \mathcal{G}$.*

2.4 Markov processes

Conditioning is a basic concept in the theory of Markov processes, which grew into a rich mathematical structure. The classical phase of the development of Markov processes may be said to have ended and is described in many books. The most classical ones are perhaps Feller (1950), Doob (1953), Chung (1967), Dynkin (1965), and more recent ones are, for example, Kingman (1972), Chung (1982), Ethier and Kurtz (1986). We recall some basic facts needed to understand the notion of Markovian networks, especially Jacksonian networks.

Consider an arbitrary state space E, with a σ-field $\mathcal{E}$. The pair $(E, \mathcal{E})$ is called a measurable state space. Let $T \subseteq R$ (typically $T = Z$, or $T = R$).

A stochastic process $(X_t, t \in T)$, with a measurable state space E, is a Markov process if

$$P(X_{t_n} \in B \mid X_{t_1}, \ldots, X_{t_{n-1}}) = P(X_{t_n} \in B \mid X_{t_{n-1}}) \text{ a.s. },$$

for all $t_1 < \ldots < t_n, n \geq 1$, and measurable B ($B \in \mathcal{E}$). This property (**Markov property**) depends only on the finite dimensional distributions of (X_t), hence it is not a property specific to some particular versions of this process.
Let $\mathcal{F}_t = \sigma(X_s, s \leq t)$, $\mathcal{F}'_t = \sigma(X_s, s \geq t)$. These two σ-fields describe, at time t, the past and the future of the process. The Markov property is equivalent to each of the following conditions

- $$P(A \cap B \mid X_t) = P(A \mid X_t)P(B \mid X_t),$$
  for $A \in \mathcal{F}_t, B \in \mathcal{F}'_t$;
- $$P(B \mid \mathcal{F}_t) = P(B \mid X_t),$$
  for $B \in \mathcal{F}'_t$;
- $$P(A \mid \mathcal{F}'_t) = P(A \mid X_t),$$
  for $A \in \mathcal{F}_t$.

An useful description of Markov processes is provided by **transition function families.**

Definition A. *A family of functions $\{P_{s,t}(x, B), s, t \in T, x \in E, B \in \mathcal{E}\}$ is a transition function family for (X_t) if*
i) $P_{s,t}(x, .)$ *is a probability measure on $(E, \mathcal{E})$, for all s, t, x;*
ii) $P_{s,t}(., B)$ *is measurable function, for all s, t, B;*
iii) $P_{s,t}(x, B) = P(X_t \in B \mid X_s = x)$ *a.s.;*
iv) $P_{s,s}(x, B) = I_A(x)$.

If (X_t) is a Markov process then its transition function family, together with some initial distribution $\mu(B) = P(X_0 \in B)$, determines this process, i.e. the finite dimensional distributions of (X_t) are determined by $\{P_{s,t}(x, B)\}$.

Lemma B. *For $s < t_1 < \ldots < t_k$,*
$$P(X_{t_1} \in B_1, \ldots, X_{t_k} \in B_k \mid X_s) = \phi(X_s) \ a.s.,$$
where
$$\phi(x) = \int_{B_1} P_{s,t_1}(x, dy_1) \int_{B_2} P_{t_1,t_2}(y_1, dy_2) \ldots \int_{B_k} P_{t_{k-1},t_k}(y_{k-1}, dy_k).$$

Proof. We apply induction. For $k = 1$, we have to prove
$$P(X_{t_1} \in B \mid X_s) = \phi(X_s) \ a.s.,$$
where
$$\phi(x) = \int_{B_1} P_{s,t_1}(x, dy_1).$$

This follows from *iii*) in Definition A.. Applying simple functions of the form

$$f(x) = \sum_{i=1}^{k} c_i I_{A_i},$$

we see immediately that we have a stronger property,

$$E(f(X_{t_1} \mid X_s) = \phi(X_s),$$

for

$$\phi(x) = \int_R P_{s,t_1}(x, dy_1) f(y_1),$$

where f is an arbitrary measurable function.
Assume now that, for a fixed n,

$$E(f_1(X_{t_1}, \ldots, f_n(X_{t_n}) \mid X_s) = \phi(X_s),$$

for

$$\phi(x) = \int_R P_{s,t_1}(x, dy_1) f_1(y_1) \ldots \int_R P_{t_{n-1},t_n}(y_{n-1}, dy_n) f_n(y_n).$$

Then

$$E(f_1(X_{t_1}, \ldots, f_{n+1}(X_{t_{n+1}}) \mid X_s) =$$
$$E(E(f_1(X_{t_1}, \ldots, f_{n+1}(X_{t_{n+1}}) \mid X_s, \ldots, X_{t_n}) \mid X_s) =$$
$$E(f_1(X_{t_1}, \ldots, f_n(X_{t_n}) E(f_{n+1}(X_{t_{n+1}}) \mid X_{t_n}) \mid X_s) =$$
$$E(f_1(X_{t_1}, \ldots, \hat{f}_n(X_{t_n}) \mid X_s),$$

where

$$\hat{f}_n(x) = f_n(x) \int_R P_{t_n,t_{n+1}}(x, dy_{n+1}) f_{n+1}(y_{n+1}).$$

Now applying the inductive assumption to $f_1, \ldots, \hat{f}_n$, we have

$$E(f_1(X_{t_1}, \ldots, f_n(X_{t_n}) \mid X_s) = \phi(X_s),$$

for an arbitrary n. Applying this formula for $f_i = I_{B_i}$, we get the assertion of the lemma.
□

Considering a Markov process (X_t), we use its transition function family to describe various properties of the process. However, it is of interest to ask whether it is possible to define a Markov process using some function families, which would form then its transition function family.

The following family of functions will be appropriate.

Definition C. *A family of functions* $\{P_{s,t}(x, B)\}$ *is a Markov transition function family if*

i) $P_{s,t}(x, .)$ *is a probability measure on* $(E, \mathcal{E})$, *for all* s, t, x;
ii) $P_{s,t}(., B)$ *is measurable function, for all* s, t, B;
iii) $P_{s,t}(x, B) = \int_E P_{s,u}(x, dy) P_{u,t}(y, B)$, *for all* $s < u < t, B \in \mathcal{E}, x \in E$;
iv) $P_{s,s}(x, B) = I_A(x)$.

The property *iii)* is called the **Chapman-Kolmogorov equation**.

The following standard theorem will be used to introduce Markov processes by given transition characteristics.

Theorem D. *If $\{P_{s,t}(x,B)\}$ is a Markov transition function family then there exists a Markov process (X_t) such that its transition function family equals to $\{P_{s,t}(x,B)\}$, and it has an arbitrary initial distribution μ.*

Countable state space

We will confine our attention to Markov processes with a countable state space, which are used frequently in applied probability models. In this case, we write $p_{ij}(s,t)$ rather than $P_{s,t}(i,\{j\})$. We consider functions $p_{ij}(s,t)$, which depend on s,t only through $h=t-s$, that is, which are **time homogeneous**. We adopt the following notation, $p_{ij}(h)=p_{ij}(s,s+h)$, and we assume that all $p_{ij}(.)$ are measurable functions. Now, from Theorem D., we have

Corollary E. *If $\{p_{ij}(h), h\geq 0, i,j\in E\}$, for a countable state space E, is a family of measurable functions fulfilling*

i) $\sum_{j\in E} p_{ij}(h)=1$, *for all* $h\geq 0, i\in E$;

ii) $p_{ij}(t)=\sum_{k\in E} p_{ik}(t-h)p_{kj}(h)$, *for all* $0\leq h\leq t, i,j\in E$;

iii) $p_{ij}(h)\geq 0, p_{ii}(0)=1$, *for all* $i,j\in E, h\geq 0$;

then there exist a Markov process $(X_t, t\geq 0)$ such that, for all $i,j\in E, h,t\geq 0$,

$$P(X_{t+h}=j \mid X_t=i)=p_{ij}(h),$$

and it has an arbitrary initial distribution μ.

We call such a processes **Markov chains in continuous time**.

For a fixed $t\geq 0$, we view $[p_{ij}(t)]_{i,j\in E}$ as an infinite matrix. We introduce additional notation for this matrix, $\mathbf{P}_t=[p_{ij}(t)]$. Suppose that, for the family $\{p_{ij}(h), i,j\in E, h\geq 0\}$, the following limits exist, and are finite:

$$q_{ij}=\lim_{t\to 0+} p_{ij}(t)/t,$$

$$q_{ii}=\lim_{t\to 0+}(p_{ii}(t)-1)/t.$$

The matrix $\mathbf{Q}=[q_{ij}]_{i,j\in E}$, is the intensity matrix of this family (or the **intensity matrix** of the corresponding Markov chain). We will also write $\mathbf{Q}=d\mathbf{P}_t/dt\,|_{t=0}$, in this case. We say that $\mathbf{Q}$ is **regular** if $\sum_{j\neq i} q_{ij}=-q_{ii}$, and $\mathbf{Q}$ is **uniform** if, in addition, $\sup_i(-q_{ii})<\infty$.

We shall consider matrices of the form $\mathbf{Q}=[q_{ij}]_{i,j\in E}$ outside the context of Markov transition function families. Such matrices which are nonnegative, and regular, we simply call then Q-matrices.

It is more convenient, in many models, to introduce processes of interest by transition intensities given by Q-matrices. This is technically possible because uniform Q-matrices uniquely determine Markov transition function families, which in turn determine Markov processes. The following classical result can be found e.g. in Feller (1968). This is a very special case of the Hille -Yosida type theorems (see e.g. Ethier and Kurtz (1986)).

Theorem F. *If* $\mathbf{Q}$ *is an uniform intensity matrix then*

$$\mathbf{P}_t = e^{\mathbf{Q}t} = \sum_{n=0}^{\infty} \mathbf{Q}^n t^n / n!$$

defines a Markov transition function family, which is the unique solution of

$$d\mathbf{P}_t/dt = \mathbf{Q}\mathbf{P}_t, \ t \geq 0,$$

and

$$d\mathbf{P}_t/dt = \mathbf{P}_t\mathbf{Q}, \ t \geq 0,$$

with the initial condition $\mathbf{P}_0$=$\mathbf{I}$ *(*$\mathbf{I}$ *is the identity matrix).*

The above differential equations are called **backward** and **forward** Kolmogorov equations, respectively. It should be clear that the forward and backward equations are not independent of each other; the solution of the backward equations with the initial conditions as above, automatically satisfies the forward equations (except for the rare situations where the solution is not unique). In the case of uniform intensity matrix $\mathbf{Q}$, neither system of equations possesses any other solutions, and hence the two systems are essentially equivalent. However, for not uniform $\mathbf{Q}$, we sometimes encounter unexpected solutions (defective solutions), for which $\sum_k p_{jk}(t) < 1$. It was shown (see e.g. Feller (1940)) that there always exists a **minimal solution** satisfying both backward and forward equations. When the minimal solution is not defective, the process is uniquely determined by either system of equations, but if it is defective, there exist infinitely many solutions satisfying the backward equations. Some of them may also satisfy the forward equations. In general, the backward equations express probabilistically meaningful conditions and lead to interesting processes, the forward equations are associated with some analytic in character assumptions, which are not probabilistically meaningful. This explains why the theory of Markov processes is better tractable in terms of semi-groups acting on functions (which correspond to the backward equations, see (1)), rather than on measures (which correspond to the forward equations, see (2)).

Corollary G. *If* $\mathbf{Q}$ *is an uniform intensity matrix then there exist a Markov process* $(X_t, t \geq 0)$ *such that* $\mathbf{Q}$ *is the intensity matrix of its transition function family, and it has an arbitrary initial distribution* μ.

A standard example of continuous time Markov chains with uniform Q-matrices is that obtained from discrete time Markov chains by a Poisson process time change.

example (Uniformization)
Let $\mathbf{P}$ be a stochastic matrix of a discrete time Markov chain $\{X(n), n \geq 1\}$. Define $\mathbf{P}_t = e^{-\lambda t}\sum_{k=0}^{\infty}(\lambda t)^k \mathbf{P}^k/k!$, i.e. $\mathbf{P}_t = e^{\mathbf{Q}t}$, for $\mathbf{Q} = \lambda(\mathbf{P} - \mathbf{Id}), \lambda > 0, t \geq 0$. Then $\mathbf{P}_t$ defines a Markov transition family with the uniform intensity matrix $\mathbf{Q}$. The corresponding (see Theorem D.) continuous time Markov chain $(X_t, t \geq 0)$, with the same initial distribution as for $\{X(n), n \geq 1\}$, is such that $X_t =^d X(N(t))$, where $\{N(t), t \geq 0\}$ is an independent Poisson process with intensity $\lambda > 0$, i.e. we can view it as a Poisson process time change, for $\{X(n), n \geq 1\}$. It is interesting that one can in a sense

reverse this argument. Namely, if $\mathbf{P}_t$ corresponds to a Markov transition function family, with an uniform intensity matrix $\mathbf{Q}$, then one defines a discrete time Markov chain $\{X(n), n \geq 1\}$, such that again $X_t =^d X(N(t))$, for the corresponding processes, with the same initial distributions. It is enough to define a stochastic matrix for this chain by $\mathbf{P} = \mathbf{Id} + \mathbf{Q}/\lambda$, where λ is an arbitrary number such that $\sup_i(-q_{ii}) \leq \lambda$. The resulting discrete time Markov chain has transitions from states to themselves. The version $(X(N(t))$, for the family corresponding to $\mathbf{P}_t$, is very convenient because its regularity of trajectories follows from the regularity of trajectories of Poisson processes. Continuous time Markov chains with uniform intensity matrices are called **uniformizable** chains. In the vast majority of applied probability models we encounter this type of chains (for more details see Keilson (1979)).

Two semigroups of operators

We revisit in this paragraph the state space $E = R$ to illustrate an elegant mathematical description of continuous time Markov processes. We start with time homogeneous Markov transition family $\{P_{s,t}(x, B), x \in R, B \in \mathcal{B}^1, 0 < s < t, s, t \geq 0\}$. For convenience we write $P(t-s, x, B)$, for $P_{s,t}(x, B)$, because of the time homogeneity assumption. Using the family $\{P(h, x, B)\}$ we define two types of operators, one acting on real measurable functions $f \to \mathbf{P}_t f$, the other one on probability measures $\mu \to \mu\mathbf{P}_t$. Fix $t \geq 0$. Define

$$(\mathbf{P}_t f)(x) = \int_R P(t, x, dy) f(y), x \in R, \tag{1}$$

$$(\mu\mathbf{P}_t)(B) = \int_R P(t, y, B)\mu(dy), B \in \mathcal{B}^1. \tag{2}$$

As a result of acting with $\mathbf{P}_t$ on a measurable function f, we obtain a measurable function $\mathbf{P}_t$f. As a result of acting with $\mathbf{P}_t$ on a probability measure μ, we obtain a probability measure $\mu\mathbf{P}_t$. In terms of the corresponding Markov chains, we have

$$(\mathbf{P}_t f)(x) = E_x(f(X(t)),$$

where $(X(t))$ is a Markov chain with the initial distribution δ_x (an atom at x), and E_x denotes the expectation taken with respect to the distribution of the process, which starts with δ_x. The measure $\mu\mathbf{P}_t$ is the distribution of X_t if the corresponding process starts with the initial distribution μ.

If we understand $\mathbf{P}_t\mathbf{P}_s$ as composition of $\mathbf{P}_t$ and $\mathbf{P}_s$ then in both cases we have (see Chapman-Kolmogorov equation)

$$\mathbf{P}_t\mathbf{P}_s = \mathbf{P}_{s+t}, \ s, t \geq 0,$$

which implies that $\{\mathbf{P}_t, t \geq 0\}$ forms two semigroups with composition, one acting on measurable functions, the second one acting on probability measures.

In the case of countable state space we have to do with the conventional matrix multiplication, and the usual meaning of matrices as operators. A measurable function f, in this case, is represented by an arbitrary row vector $(f(i))_{i \in E}$, a probability measure μ is represented by an arbitrary probability vector $(\mu(i))_{i \in E}$. We have

$$(\mathbf{P}_t f^T)(i) = \sum_{k \in E} p_{ik}(t) f(k),$$

$$(\mu\mathbf{P}_t)(i) = \sum_{k\in E} \mu(k)p_{ki}(t), t \geq 0.$$

Stationary processes, invariant measures

Let the index set T be, in this paragraph, R, R_+, or N.

A stochastic process $(X_t, t \in T)$ is (strictly) **stationary** if

$$P(X_{t_1+h} \in B_1, \ldots, X_{t_n+h} \in B_n) = P(X_{t_1} \in B_1, \ldots, X_{t_n} \in B_n),$$

for all $t_1 < \ldots < t_n, h \geq 0, n \geq 0, B_i \in \mathcal{E}$.
This is a property of finite dimensional distributions, i.e. it is not dependent on particular versions of processes.

We will be interested in this property for continuous time Markov chains.

Let $\{\mathbf{P}_t, t \geq 0\}$ be a family of matrices corresponding to a Markov transition function family with a countable state space. A probability measure $\boldsymbol{\pi}$ is **invariant** (stationary) with respect to $\{\mathbf{P}_t\}$ if

$$\boldsymbol{\pi}\mathbf{P}_t = \boldsymbol{\pi},$$

for all $t \geq 0$.

Theorem H. *Suppose $\boldsymbol{\pi}$ is invariant with respect to $\{\mathbf{P}_t, t \geq 0\}$. Then the corresponding Markov chain $(X_t, t \geq 0)$, which starts with the initial distribution $\boldsymbol{\pi}$, is stationary.*

Proof. It is immediate from Lemma B. □

In the case of uniformizable Markov chains, we have a condition for invariance in terms of intensity matrices.

Lemma I. *Suppose that for $\{\mathbf{P}_t\}$ the corresponding intensity matrix $\mathbf{Q}$ is uniform then*

$$\boldsymbol{\pi}\mathbf{Q} = \mathbf{0}$$

if and only if $\boldsymbol{\pi}$ is invariant with respect to $\{\mathbf{P}_t\}$.

Proof. From Theorem F. we have $\mathbf{P}_t = \sum_{k=0}^{\infty} \mathbf{Q}^k t^k/k!$, hence $\boldsymbol{\pi}\mathbf{P}_t = \sum_{k=0}^{\infty} \boldsymbol{\pi}\mathbf{Q}^k t^k/k! = \boldsymbol{\pi}\mathbf{Q}^0 = \boldsymbol{\pi}\mathbf{Id} = \boldsymbol{\pi}$. □

We say in this case that $\boldsymbol{\pi}$ is **invariant with respect to Q**.

In general it is not a simple task to find an invariant measure. However the problem of existence of such measures is a classical one and is resolved in a broad generality with use of **ergodic theorems** (see Doob (1953), Feller (1968), Chung (1967), Kingman (1972), Asmussen (1987), and for a recent treatment Meyn and Tweedie (1993 a,b)).

We recall two classical results on stationary probabilities.

Theorem J. *Consider a Markov chain $\{X(n), n \geq 0\}$ with a finite state space $\mathcal{S}$, and transition probability matrix $\mathbf{P} = [p_{ij}]$ ($\mathbf{P}^k = [p_{ij}^{(k)}]$)). If for some m, $\min_{i,j\in\mathcal{S}} p_{ij}^{(m)} = \epsilon > 0$ then*

(i) $\pi_j = \lim_{k\to\infty} p_{ij}^{(k)}$ *exists for all i, j, and is independent of i's;*

(ii) $\sum_i \pi_i p_{ij} = \pi_j$, *i.e.* $\pi = (\pi_j)$ *is a stationary (invariant) measure.*

Proof. We adopt Liggett's (1985) coupling proof.
Let $\mathbf{X}(n) = (X^{(1)}(n),\ X^{(2)}(n))$ be a Markov chain with the state space $\mathcal{S} \times \mathcal{S}$, and the transition probabilities

$$p_{(i,j)(k,l)} = \begin{cases} p_{ik}p_{jl} & for \quad i \neq j \\ p_{ik} & for \quad i = j,\ k = l \\ 0 & for \quad i = j,\ k \neq l \end{cases}$$

a) $(X^{(1)}(n))$ and $(X^{(2)}(n))$ are Markov chains with the transition probability matrix $\mathbf{P}$.
b) $\lim_{n\to\infty} P_{(i,j)}(X^{(1)}(n) = X^{(2)}(n)) = 1$, where $P_{(i,j)}(\cdot) = \mathbf{Pr}(\cdot \mid \mathbf{X}(0) = (i,j))$, for all $(i,j) \in \mathcal{S} \times \mathcal{S}$. Indeed, for m, for which $\min_{i,j\in\mathcal{S}} p_{ij}^{(m)} = \epsilon > 0$, $P_{(i,j)}(X^{(1)}(m) = X^{(2)}(m)) \geq \sum_{s\in\mathcal{S}} p_{is}^{(m)} p_{js}^{(m)} \geq \epsilon$. Thus $P_{(i,j)}(X^{(1)}(m) \neq X^{(2)}(m)) \leq 1-\epsilon$, and by the Markov property $P_{(i,j)}(X^{(1)}(nm) \neq X^{(2)}(nm)) \leq (1-\epsilon)^n$, which gives b).
c) $\lim_{k\to\infty} \mid p_{ij}^{(k)} - p_{sj}^{(k)} \mid = 0$. Indeed, $\mid p_{ij}^{(k)} - p_{sj}^{(k)} \mid = \mid P_{(i,s)}(X^{(1)} = j) - P_{(i,s)}(X^{(2)} = j) \mid \leq P_{(i,s)}(X^{(1)} \neq X^{(2)}) \to_{k\to\infty} 0$.
d) $M_j(k) = \max_{i\in\mathcal{S}} p_{ij}^{(k)}$ is a nonincreasing function of k for all $j \in \mathcal{S}$, $m_j(k) = \min_{i\in\mathcal{S}} p_{ij}^{(k)}$ is a nondecreasing function of k for all $j \in \mathcal{S}$. We have

$$M_j(k+1) = \max_i \sum_{s\in\mathcal{S}} p_{is} p_{sj}^{(k)} \leq \max_i \sum_s p_{is} M_j(k) = M_j(k),$$

and likewise $m_j(k+1) \geq m_j(k)$. Thus there exist the limits $\lim_k M_j(k) = M_j$ and $\lim_k m_j(k) = m_j$. In order to prove (i) it is enough to show that $M_j = m_j$. Now

$$\mid M_j - m_j \mid \leq \mid M_j - p_{ij}^{(k)} \mid + \mid p_{ij}^{(k)} - p_{sj}^{(k)} \mid + \mid m_j - p_{sj}^{(k)} \mid,$$

where the terms on the right hand of the above inequality can be made arbitrary small from c) (middle) and d).
The validity of (ii) is a consequence of (i) and the Chapman-Kolmogoroff equations. □

For continuous time Markov processes with a general state space we have the following

Theorem K. (Sevastyanov (1957)) *Suppose* (X_t) *is a Markov process with a time homogeneous Markov transition family* $\{P(t,x,B), x \in E,\ t > 0,\ B \in \mathcal{B}(E)\}$, *where* E *is a measurable space with Borel field* $\mathcal{B}(E)$. *If for any* $\epsilon > 0$, *there exist a measurable set* C, *a probability measure* R *on* E, *and* $s, k, K > 0$ *such that*

(i) $kR(A) \leq P(s,x,A)$, *for* $x \in C$, $A \subseteq C$;

(ii) *for any initial distribution, there exist* t_0 *such that* $\mathbf{Pr}(X_t \in C) \geq 1-\epsilon$, $t \geq t_0$;

(iii) $\mathbf{Pr}(X_t \in A) \leq KR(A) + \epsilon$ *for* $A \subseteq C$, *and* $t \geq t_0$,

then (X_t) *has a unique stationary (invariant) distribution* π, *such that for every initial conditions* $\mathbf{Pr}(X_t \in \cdot)$ *converges in variation to* $\pi(\cdot)$ *when* $t \to \infty$.

Time reverse processes are helpful in determining particular forms of invariant measures.

Suppose $(X_t, t \geq 0)$ is a stationary continuous time stochastic process. The **time reverse process** of (X_t) is a stochastic process $(X_t^{\leftarrow}, t \geq 0)$, with the following finite dimensional distributions

$$P(X_{t_1}^{\leftarrow} \in B_1, \ldots, X_{t_n}^{\leftarrow} \in B_n) = P(X_{S-t_1} \in B_1, \ldots, X_{S-t_n} \in B_n),$$

for all $0 \leq t_1 \leq \ldots \leq t_n \leq S, n \geq 1, B_i \in \mathcal{E}, i = 1, \ldots, n$.

Remarks.
1. If $(X_t, t \in R)$ is stationary then $X_t^{\leftarrow} = X_{-t}$ is the time reverse process of $(X_t, t \geq 0)$.

2. If $(X_t, t \geq 0)$ is a stationary Markov chain in continuous time, with the invariant measure $\boldsymbol{\pi}$ then the formula

$$q'_{ij} = \frac{\pi_j}{\pi_i} q_{ji}, \; i, j \in E, \tag{3}$$

defines an intensity matrix $\mathbf{Q}'$, which is uniform whenever $(X_t, t \geq 0)$ is uniformizable. The corresponding (see Corollary G.) to $\mathbf{Q}'$ Markov chain, which starts with the initial measure $\boldsymbol{\pi}$, is the time reverse process of $(X_t, t \geq 0)$.

In a number of cases one is able to verify that a matrix of the form (3) is an intensity matrix, for a given $\mathbf{Q}$, and a given probability vector $\boldsymbol{\pi}$ (which is not assumed to be invariant with respect to $\mathbf{Q}$). In such cases an important role in finding invariant measures for many processes (encountered for example in queueing theory) plays the following theorem.

Theorem L. *Suppose $\mathbf{Q}$ is a uniform intensity matrix, and $\boldsymbol{\pi}$ is a positive probability vector, such that $\mathbf{Q}'$ given by (3) is an intensity matrix then*

$$\boldsymbol{\pi}\mathbf{Q} = 0,$$

and

$$\boldsymbol{\pi}\mathbf{Q}' = 0.$$

Proof. We verify that $\boldsymbol{\pi}\mathbf{Q} = 0$. From (3)

$$\sum_{j \in E} \pi_j q_{jk} = \pi_k \sum_{j \in E} q'_{kj},$$

but since we assumed that $\mathbf{Q}'$ is an intensity matrix, we have $\sum_j q'_{kj} = 0,$, and hence

$$\boldsymbol{\pi}\mathbf{Q} = 0.$$

To check that $\boldsymbol{\pi}\mathbf{Q}' = 0$, write

$$\sum_{j \in E} \pi_j q'_{jk} = \pi_k \sum_{j \in E} q_{kj} = 0,$$

since $\mathbf{Q}$ is an intensity matrix (and as such regular). □

Corollary M. *Suppose $(X_t, t \geq 0)$ is a Markov chain with the corresponding intensity matrix* $\mathbf{Q}$*, and π is a positive initial probability measure, such that $\mathbf{Q}'$ given by (3) is an intensity matrix then π is the stationary measure for $(X_t, t \geq 0)$, and π is the stationary measure for the time reverse process of $(X_t, t \geq 0)$ ($\mathbf{Q}'$ is the intensity matrix for the time reverse process of $(X_t, t \geq 0)$).*

A stationary process $(X_t, t \geq 0)$ is **time reversible** if for the time reverse process $(X_t^{\leftarrow}) =^d (X_t)$.

From Theorem L. we see that if $\pi_j q_{jk} = \pi_k q_{kj}$, $j, k \in E, j \neq k$, for a probability measure π then the corresponding time continuous stationary Markov chain is time reversible, and π is an invariant measure for it.

2.5 Point processes on R, martingales

The theory of point processes has been developed along with applications in numerous fields in the last three decades. Palm's (1943) paper in telecommunications is a pioneering work in modeling flows of customers in queues as point processes. A mathematical foundation for such an approach was given by Khintchine (1960) and Ryll-Nardzewski (1961). The theory of point processes is intimately connected with the subject of measure and integration (a point process is a random measure), however many results can be understood without a deep knowledge of measure theory, and can be very helpful in understanding results in the applied probability literature. The basic monographs in this field are Daley and Vere-Jones (1988) and Kallenberg (1983), less formal approach can be found in Cox and Isham (1980), Snyder and Miller (1991) and Grandell (1976). Applications of point processes in queueing are discussed by Bremaud (1981), Franken et al. (1981) and Rolski (1981). A very good introduction to point processes in applied probability one can find in Serfozo (1990). In this section we recall some basic notions and results from point processes on R , sufficient to understand the later results on stochastic ordering and Poissonian flows in networks.

We start with introducing more structure on probability spaces by defining some useful σ-fields.

Let $(\Omega, \mathcal{F}, P)$ be a probability space . A **history** (or **filtration**) is a family of σ-fields $(\mathcal{F}_t, t \geq 0)$ such that for all $s \leq t, \mathcal{F}_s \subseteq \mathcal{F}_t \subseteq \mathcal{F}$.
If $\mathcal{F}_t = \cap_{h>0} \mathcal{F}_{t+h}$ then we say that the history $(\mathcal{F}_t)$ is right-continuous. A typical example of a history is $\mathcal{F}_t^X = \sigma(X_s, 0 \leq s \leq t)$, for a given stochastic process $(X_t, t \geq 0)$ on $(\Omega, \mathcal{F}, P)$, i.e. the **internal history** of (X_t). A process (X_t) is **adapted** to a history $(\mathcal{F}_t)$ if $\mathcal{F}_t^X \subseteq \mathcal{F}_t$, for all t.

If we consider a stochastic process $(X(t), t \geq 0)$ as a function of two variables $(t, \omega) \to X_t(\omega)$, it is natural to require some regularity conditions on this function. The domain of this function is the product $[0, \infty) \times \Omega$ in which we define the product σ-field denoted by $\mathcal{B}^1 \otimes \mathcal{F} = \sigma(B \times F, B \in \mathcal{B}^1, F \in \mathcal{F})$. If the function $(t, \omega) \to X_t(\omega)$ is $\mathcal{B}^1 \otimes \mathcal{F}$-measurable we say that the process (X_t) is **measurable**. But having a history $(\mathcal{F}_t)$ we can introduce a more delicate concept, by considering the process (X_t) on $[0, t]$ instead on $[0, \infty)$, for each fixed $t \geq 0$. Denote the Borel σ-field on $[0, t]$ by $\mathcal{B}_{[0,t]}$. If for each

$t \geq 0$, the function $(t,\omega) \to X_t(\omega)$ with the domain $[0,t] \times \Omega$ is $\mathcal{B}_{[0,t]} \otimes \mathcal{F}_t$-measurable , we say that (X_t) is **progressive** with respect to $(\mathcal{F}_t)$. This property is dependent on particular versions of (X_t), and is connected with regularity properties for trajectories of the process (X_t). This is evident from the following lemma.

Lemma A. *If (X_t) is adapted to $(\mathcal{F}_t)$ and almost all trajectories of (X_t) are right-continuous (or left-continuous) then (X_t) is progressive.*

Proof. Fix $t_0 \geq 0$, and define for $n \in N, k = 1, \ldots, 2^n$, and $t \leq t_0 : X_0^n = X_0$,

$$X_t^n(\omega) = \sum_{k=1}^{2^n} X_{k2^{-n}t_0}(\omega) I_{[(k-1)2^{-n}t_0, k2^{-n}t_0))}(t).$$

Then for all a, $s \leq t_0$,

$$\{(s,\omega) : X_s^n(\omega) \leq a\} = \cup_{k=1}^{2^n}[(k-1)2^{-n}t_0, k2^{-n}t_0) \times \{\omega : X_{k2^{-n}t_0}(\omega) \leq a\}$$

which of course belongs to $\mathcal{B}_{[0,t_0]} \times \mathcal{F}_{t_0}$. If X_t is right-continuous, $X_t(\omega)$ is the limit of $X_t^n(\omega)$ for all $(t,\omega) \in [0,t_0] \times \Omega$, therefore $(t,\omega) \to X_t(\omega)$ is $\mathcal{B}_{[0,t_0]} \otimes \mathcal{F}_{t_0}$ - measurable. Since t_0 was arbitrary, (X_t) is progressive. □

A similar argument can be used for processes with left-continuous trajectories, however processes with left-continuous trajectories are measurable in a stronger sense.

If for a process $(X_t, t \geq 0)$ the function $(t,\omega) \to X_t(\omega)$ is measurable with respect to

$$\sigma((s,\infty) \times A : s \geq 0, A \in \mathcal{F}_s)$$

then (X_t) is called **predictable** with respect to $(\mathcal{F}_t)$.

Lemma B. *If (X_t) is adapted to $(\mathcal{F}_t)$ and almost all trajectories of (X_t) are left-continuous then (X_t) is predictable.*

Proof. The proof is through approximation by simple processes as in the case of progressive processes (for details see e.g. Bremaud (1981), T5 in I). □

We will use the notions of progressivity and predictability in the process of introducing stochastic intensities for point processes.

Another classical concept we recall for use in this paragraph is that of stopping times.

A random variable S is an $(\mathcal{F}_t)$ **stopping time** if for each t

$$\{\omega : S(\omega) \leq t\} \in \mathcal{F}_t.$$

We summarize some useful properties of stopping times in the following lemma.

Lemma C. *Let $(\mathcal{F}_t, t \geq 0)$ be a history on a probability space $(\Omega, \mathcal{F}, P)$. Then*

- *each real number $t \geq 0$ is an $(\mathcal{F}_t)$ stopping time;*
- *if S,T are $(\mathcal{F}_t)$ stopping times then $S+T, \min(S,T), \max(S,T)$ are $(\mathcal{F}_t)$ stopping times ;*

- *if* $\{S_n\}_{n\geq 1}$ *is a sequence of* $(\mathcal{F}_t)$ *stopping times then* $\sup_n S_n, \inf_n S_n$ *are* $(\mathcal{F}_t)$ *stopping times.*

In order to describe the history up to a random time (or just before a random time) we introduce in addition the following σ-fields.

Let $(\mathcal{F}_t)$ be a history on $(\Omega, \mathcal{F}, P)$, and S an $(\mathcal{F}_t)$ stopping time. The **past at time** S is the σ-field

$$\mathcal{F}_S = \sigma\{A : A \cap \{S \leq t\} \in \mathcal{F}_t, t \geq 0\},$$

the **strict past** at time S is the σ-field

$$\mathcal{F}_{S-} = \sigma\{A \cap \{\omega : S(\omega) > t\}, A_0 : A \in \mathcal{F}_t, A_0 \in \mathcal{F}_0, t \geq 0\}.$$

In general the past σ-fields and stopping times behave as expected.

Lemma D. *If* S, T *are* $(\mathcal{F}_t)$ *stopping times then*

- $\mathcal{F}_{S-} \subseteq \mathcal{F}_S$ *and* S *is* $\mathcal{F}_{S-}$ *measurable;*
- *if* $S \leq T$ *then* $\mathcal{F}_S \subseteq \mathcal{F}_T$, *and* $\mathcal{F}_{S-} \subseteq \mathcal{F}_{T-}$.

Stochastic processes stopped at random times behave also in a regular way. Before we formulate a result, recall that an $(\mathcal{F}_t)$ stopping time S is **predictable** if there exist strictly increasing sequence of $(\mathcal{F}_t)$ stopping times $\{S_n\}_{n\geq 1}$, such that $\lim_n S_n = S$ and $S_n < S$ on the set $\{\omega : 0 < S(\omega) < \infty\}$. We say that $\{S_n\}$ foretells S. The following theorem can be found in Chung (1982).

Theorem E. *Suppose* $(\mathcal{F}_t, t \geq 0)$ *is a right-continuous history on* $(\Omega, \mathcal{F}, P)$, *and* S *is a finite* $(\mathcal{F}_t)$ *stopping time.*

- *If* (X_t) *is progressive then the random variable* X_S , *defined as* $X_{S(\omega)}(\omega)$ *for* $\omega \in \Omega$, *is* $\mathcal{F}_S$ *measurable;*
- *If* (X_t) *is predictable and* S *is predictable then* X_{S-} *is* $\mathcal{F}_{S-}$ *measurable.*

For a sequence of stopping times we define a counting process as follows.

Let $(\mathcal{F}_t, t \geq 0)$ be a right-continuous history on a probability space $(\Omega, \mathcal{F}, P)$. For a strictly increasing sequence $\{\tau_n\}_{n\geq 1}$ of $(\mathcal{F}_t)$ stopping times the **counting process** is

$$N_t(\omega) = \sum_{n=1}^{\infty} I_{\{(t,\omega):\tau_n(\omega)\leq t\}}(t, \omega), \tag{1}$$

where I denotes the indicator function, $t \geq 0$.

We usually omit ω in the above notation, and write N_t or $N(t)$.

We will refer to (N_t) also as to a **point process**, because each trajectory of it determines a sequence of points on R_+. If almost all trajectories are such that $\tau_1 < \tau_2 < \ldots$, and $\tau_n \to_n \infty$, then we say that this point process is **non-explosive.** The sequence of stopping times $\{\tau_n\}_{n\geq 1}$ will be also called sometimes a point process. Another- measure theoretic- approach to point processes on R is given in Ryll-Nardzewski (1961), we will use this approach later on for stochastic ordering of point processes.

If we do not introduce an additional structure on the probability space of interest by histories, we can define a counting process for an arbitrary sequence of increasing random variables. A connection with the above structural approach is given by the following lemma.

Lemma F. *Suppose that on the probability space $(\Omega, \mathcal{F}, P)$ is given a sequence of random variables $\tau_1 < \tau_2 < \dots$. Consider the corresponding counting process (N_t) given by (1), and the internal history of (N_t), $\mathcal{F}_t^N = \sigma(N_s, s \leq t)$ then*

- *for each n, τ_n is an $(\mathcal{F}_t^N)$ stopping time;*
- $\sigma(\tau_1, \dots, \tau_n) = \mathcal{F}_{\tau_n}^N$;
- $\mathcal{F}_t^N = \sigma(I_{\{\tau_n \leq s\}} : 0 \leq s \leq t, n \geq 1)$.

Proof. See e.g. $T23$ in A2 of Bremaud (1981). □

Probability spaces with a rich structure of histories are introduced to characterize point processes via martingales. Let us recall the definition of a martingale and some basic examples of martingales. The time index in the following definition can be discrete or continuous.

Let $(\mathcal{F}_t, t \geq 0)$ be a history on a probability space $(\Omega, \mathcal{F}, P)$. A stochastic process $(X_t, t \geq 0)$, adapted to $(\mathcal{F}_t)$, is $(\mathcal{F}_t)$ **martingale** if

- $E \mid X_t \mid < \infty, \; t \geq 0$;
- $E(X_t \mid \mathcal{F}_s) = X_s$ a.s., for all $s \leq$t.

Examples.

1. Let $\{X_i\}_{i\geq 1}$ be i.i.d. such that EX=0, and $\mathcal{F}_n^X = \sigma(X_1, \dots, X_n)$. Then $S_n = X_1 + \dots + X_n$ is $(\mathcal{F}_n^X)$ martingale.

2. Let $\{X_i\}_{i\geq 1}$ be i.i.d. such that EX=1. Then $P_n = X_1 \dots X_n$ is $(\mathcal{F}_n^X)$ martingale.

3. Let $(\mathcal{F}_t)$ be a history on a probability space $(\Omega, \mathcal{F}, P)$ and X a random variable such that $E \mid X \mid < \infty$. Then $X_t = E(X \mid \mathcal{F}_t)$ forms an $(\mathcal{F}_t)$ martingale.

4. Let (X_t) be a process with independent increments such that $X_0 = 0$, and EX$_t = 0$. Then (X_t) is $(\mathcal{F}_t^X)$ martingale. If in addition EX$_t^2 < \infty, t \geq 0$, then $Y_t = X_t^2$-EX$_t^2$ is $(\mathcal{F}_t^X)$ martingale.

5. Suppose $(X_t, t \geq 0)$ is a continuous time Markov chain with an intensity matrix $\mathbf{Q}$. Then

$$M_t = \sum_{s \leq t} f(X_{s-}, X_s) - \int_0^t \sum_{j \neq X_s} q_{X_s, j} f(X_s, j) ds, \tag{2}$$

is $(\mathcal{F}_t^X)$ martingale, where f is a nonnegative real function such that $f(i,i) = 0, i \in E$, and X_{s-} denotes the left-hand limit value.

Intuitively speaking martingales are processes with a constant expected value over time, but which become more variable with the passing of time. Such intuitions are expressed more formally by the following theorem of Strassen (1965).

Theorem G. *If $(X_t, t \geq 0)$ is $(\mathcal{F}_t^X)$ martingale then $X_s <_{icx} X_t$ for all $s \leq t$. Conversely, if $\{\mu_t, t \geq 0\}$ is a family of distributions such that $\mu_s <_{icx} \mu_t$ for all $s \leq t$, and the expected values $\int_R x d\mu_t(x)$ are constant with respect to t then there exist a process (X_t) such that X_t has distribution μ_t, $t \geq 0$, and (X_t) is $(\mathcal{F}_t^X)$ martingale.*

It is clear that processes with strictly increasing trajectories can not be martingales, because they do not have a constant expected value over time. However they can sometimes be compensated by other processes to produce martingales. This is a common situation for counting processes for which compensating processes (compensators) have some additional intuitive meaning connected with so called stochastic intensities of point processes. Our task now is to introduce and characterize stochastic intensities for point processes which will be useful later for stochastic ordering and Poissonian flows in networks. Before we go into a general context we recall some related facts about Poisson processes (see Bremaud (1981)).

Let $(\mathcal{F}_t, t \geq 0)$ be a history on $(\Omega, \mathcal{F}, P)$ and (N_t) a counting process adapted to $(\mathcal{F}_t)$. Suppose (λ_t) is a nonnegative stochastic process on $(\Omega, \mathcal{F}, P)$ which is $\mathcal{F}_0$ measurable and locally integrable i.e., $\int_0^t \lambda_s ds < \infty$ a.s. for all $t \geq 0$. (N_t) is a **doubly stochastic $(\mathcal{F}_t)$ Poisson process** if

$$E(\xi^{(N_t - N_s)} \mid \mathcal{F}_s) = \exp\{(\xi - 1)\int_s^t \lambda_u du\} \text{ a.s.} \tag{3}$$

for all $s \leq$t.

The equation (3) expresses in terms of generating functions that, conditionally on $\mathcal{F}_0$, (N_t) has independent increments , and the increment $N_t - N_s$ has Poisson distribution with the parameter $\int_s^t \lambda_u$ du. If (λ_t) is a deterministic function then (N_t) is a (**nonhomogeneous**) $(\mathcal{F}_t)$ Poisson process. In the special case when $(\mathcal{F}_t) = (\mathcal{F}_t^N)$, and $\lambda_t = 1, t \geq 0$ $(\lambda_t = \lambda, t \geq 0)$, (N_t) is a **standard (stationary)** Poisson process. If (N_t) is $(\mathcal{F}_t)$ Poisson process then (N_t) is $(\mathcal{F}_t^N)$ Poisson process, but the converse statement is not true (some examples will be given for queues). If we do not emphasize with respect to which history (N_t) is Poisson, then the internal history $(\mathcal{F}_t^N)$ is taken in the definition.

Note that the above introduced doubly stochastic $(\mathcal{F}_t)$ Poisson process can be compensated to give an $(\mathcal{F}_t)$ martingale. Indeed from (3) we have

$$E(N_t - N_s \mid \mathcal{F}_s) = E(\int_s^t \lambda_u du \mid \mathcal{F}_s),$$

which implies that

$$M_t = N_t - \int_0^t \lambda_u du \tag{4}$$

is an $(\mathcal{F}_t)$ martingale, provided M_t is integrable. The process given by $A_t = \int_0^t \lambda_u du, t \geq 0$, is called $(\mathcal{F}_t)$ **compensator** of (N_t). The process (λ_t) is called an $(\mathcal{F}_t)$ **stochastic intensity** of (N_t). A technical role of this stochastic intensity is that it can be used to compute some integrals with respect to the trajectories of (N_t).

Lemma H. *Let $(\mathcal{F}_t)$ be a history on $(\Omega, \mathcal{F}, P)$. If (N_t) is a doubly stochastic $(\mathcal{F}_t)$ Poisson process with the $(\mathcal{F}_t)$ stochastic intensity (λ_t) then*

$$E(\int_0^t X_u dN_u) = E(\int_0^t X_u \lambda_u du), \tag{5}$$

for all (X_t) which are predictable and nonnegative.

Proof. This is a direct consequence of the fact that (M_t) defined above is an $(\mathcal{F}_t)$ martingale if we take simple processes (X_t). For arbitrary processes we use a standard approximation method. □

An important link between martingales and point processes is the following theorem of Watanabe (1964)

Theorem I. *Let $(\mathcal{F}_t)$ be a history on $(\Omega, \mathcal{F}, P)$, and (N_t) a counting process adapted to $(\mathcal{F}_t)$. Suppose that $\lambda(t)$ is a nonnegative, locally integrable function. Then (N_t) is a nonhomogeneous $(\mathcal{F}_t)$ Poisson process if and only if (M_t) given by (4) is $(\mathcal{F}_t)$ martingale.*

Following Bremaud (1981) we define a stochastic intensity in the following way.

Let $(\mathcal{F}_t)$ be a history on $(\Omega, \mathcal{F}, P)$, and (N_t) a counting process adapted to $(\mathcal{F}_t)$. Suppose that (λ_t) is a nonnegative progressive process. We say that (λ_t) is an $(\mathcal{F}_t)$ **stochastic intensity** of (N_t) if (λ_t) is locally integrable and the condition (5) is fulfilled

An alternative characterization of stochastic intensities is given in the following theorem.

Theorem J. *Let $(\mathcal{F}_t)$ be a history on $(\Omega, \mathcal{F}, P)$, and (N_t) a counting process adapted to $(\mathcal{F}_t)$. Suppose that (λ_t) is a nonnegative progressive process. Then (λ_t) is an $(\mathcal{F}_t)$ stochastic intensity of (N_t) if and only if (N_t) is nonexplosive and*

$$M_t^{loc} = N_{t\wedge\tau_n} - \int_0^{t\wedge\tau_n} \lambda_u du,$$

is $(\mathcal{F}_t)$ martingale for all $n \geq 1$, where $\{\tau_n\}$ are the jumps of (N_t).

If the condition above is fulfilled with a nonexplosive $\{\tau_n\}$, we say that (M_t) given by (4) is a local martingale with the localizing sequence $\{\tau_n\}$. From the above theorem we have

$$E(N_t - N_s \mid \mathcal{F}_s) = E(\int_s^t \lambda_u du \mid \mathcal{F}_s),$$

and if (λ_t) is right continuous and bounded it follows that

$$\lim_{h\to 0}(1/h)E(N_{s+h} - N_s \mid \mathcal{F}_s) = \lambda_s \text{ a.s.}$$

This explains the use of the name "stochastic intensity".

The main point of interest, regarding stochastic intensities, will be now a regenerative formula for stochastic intensities.

Consider on a probability space $(\Omega, \mathcal{F}, P)$ a sequence of random variables $0 < \tau_1 < \tau_2 < \ldots$, such that $\lim_n \tau_n = \infty$ (put $\tau_0 = 0$). Let (N_t) be the corresponding counting process, and $(\mathcal{F}_t^N)$ its internal history. Assume that the conditional distribution $P(\tau_{n+1} - \tau_n \leq x \mid \tau_0, \ldots, \tau_n), n \geq 0$, has a density with respect to the Lebesgue measure, for each $\omega \in \Omega$, i.e.

$$P(\tau_{n+1} - \tau_n \leq x \mid \mathcal{F}_{\tau_n}^N)(\omega) = \int_0^x f^{(n+1)}(u, \omega)du,$$

where the function $(u, \omega) \to f^{(n)}(u, \omega)$ is $\mathcal{B}^1 \otimes \mathcal{F}_{\tau_n}^N$ measurable.

Theorem K. *Under the assumptions above the process $(\lambda_t, t \geq 0)$ defined by*

$$\lambda_t = \frac{f^{(n+1)}(t - \tau_n)}{1 - \int_0^{t-\tau_n} f^{(n+1)}(u)du}, \text{ for } t \in [\tau_n, \tau_{n+1})$$

is an $(\mathcal{F}_t^N)$ stochastic intensity of (N_t).

For the proof we need some technical lemmas.

Lemma L. *Let $(\mathcal{F}_t)$ be a history on $(\Omega, \mathcal{F}, P)$, and (X_t) be progressive. If, for each bounded stopping time S, X_S is integrable and $E(X_S) = E(X_0)$ then (X_t) is an $(\mathcal{F}_t)$ martingale.*

Proof. For $0 \leq s \leq t$ and $A \in \mathcal{F}_s$ define $T(\omega) = t$; $S(\omega) = sI_A(\omega) + tI_{A^c}(\omega)$. It is clear that T and S are bounded stopping times, so from our assumption we have $EX_S = EX_T$, which is

$$E(X_s I_A) + E(X_t I_{A^c}) = EX_t.$$

This implies that $E(X_s I_A) = E(X_t I_A)$, which from Definition C. gives $E(X_t \mid \mathcal{F}_s) = X_s$. □

Lemma M. *Suppose (X_t) is a stochastic process. If $A \in \sigma(X_u : u \in (s,t))$, for some $s < t$ then $A \in \sigma(X_u : u \in I)$, where I is a countable set and $I \subseteq (s,t)$.*

Proof. Consider the following class

$$\mathcal{A} = \{A : A \in \sigma(X_u : u \in I), \text{for some countable} I \subseteq (s,t)\}.$$

Of course sets $\{\omega : X_u(\omega) \in B\}$, for $u \in (s,t)$, and $B \in \mathcal{B}^1$, belong to $\mathcal{A}$. It is clear that $\mathcal{A}$ is a σ-field. Because $\mathcal{A}$ is a σ-field containing all $\sigma(X_u), u \in (s,t)$, it contains also $\sigma(X_u : u \in (s,t))$. This completes the proof. □

Lemma N. *Let $(N_t, t \geq 0)$ be a counting process , and $(\mathcal{F}_t^N)$ its internal history. If S is an $(\mathcal{F}_t^N)$ stopping time then*

$$\mathcal{F}_S^N = \sigma(N_{t\wedge S}, t \geq 0).$$

Proof. Recall that $\mathcal{F}_S^N = \sigma\{A : A \cap \{S \leq t\} \in \mathcal{F}_t^N, t \geq 0\}$. To prove "$\supseteq$", take for an arbitrary u , and $k \geq 0$ the set $\{N_{u\wedge S} = k\}$. It is enough to show that $\{N_{u\wedge S} = k\} \cap \{S \leq t\} \in \mathcal{F}_t^N$. If $t \leq u$ than it is equal to $\{N_S = k\} \cap \{S \leq t\}$. Because N_t is right-continuous $\{N_S = k\} \cap \{S \leq t\} = \cup_{s \leq t}(\{N_s = k\} \cap \{S = s\})$, where s are rational numbers. Of course each $\{N_s = k\} \cap \{S = s\}$ belongs to $\mathcal{F}_s^N$, and hence to $\mathcal{F}_t^N$. The case $u \leq t$ is similar. To prove "$\subseteq$" , assume that S is a simple random variable of the form

$$S = \sum_{i=1}^{\infty} a_i I_{B_i},$$

for some $0 \leq a_1 \leq a_2 \leq \ldots$. Then for an arbitrary $A \in \mathcal{F}_S^N$ we have a decomposition

$$A = \cup_{i=1}^{\infty} A_i,$$

where $A_i = A \cap B_i = A \cap \{S = a_i\} \in \mathcal{F}_{a_i}^N$. From this I_{A_i} is $\mathcal{F}_{a_i}^N$ measurable, and from Lemma M. we deduce that

$$A_i \in \sigma(N_u : u \in I),$$

for a countable set $I \subseteq [0, a_i]$. Now if $\omega \in A_i$ then $S = a_i$, hence for $u \in I,\ u \wedge S = u$, and we have

$$A_i \in \sigma(N_{u\wedge S} : u \in I).$$

From this $A \in \sigma(N_{t\wedge S}, t \geq 0)$. For general S we define, for $k \geq 1$

$$S_k = \sum_{i \geq 1} (i/2^k) I_{((i-1)/2^k \leq S < (i/2^k))}.$$

Because $S_k \geq S$ we have $\mathcal{F}_S^N \subseteq \mathcal{F}_{S_k}^N = \sigma(N_{t\wedge S_k}, t \geq 0)$. Define

$$C_k = \{\omega : N_t(\omega) = N_{S(\omega)}(\omega),\ \text{for } t \in [S, S + 1/2^k]\}.$$

Then $\mathcal{F}_S^N \cap C_k \subseteq \sigma(N_{t\wedge S_k}, t \geq 0) \cap C_k$, for all $k \geq 1$. On C_k, $N_{t\wedge S_k} = N_{t\wedge S}$, hence

$$\mathcal{F}_S^N \cap C_k \subseteq \sigma(N_{t\wedge S}, t \geq 0) \cap C_k,$$

for all $k \geq 1$.
Now because (N_t) is right-continuous the sequence $\{C_k\}$ is increasing to Ω. This implies that $\mathcal{F}_S^N \subseteq \sigma(N_{t\wedge S}, t \geq 0)$. □

Lemma O. *Let S be an $(\mathcal{F}_t^N)$ stopping time. Then there exist a sequence $\{R_n\}_{n\geq 1}$ of $\mathcal{F}_{\tau_n}^N$ measurable random variables, such that on the set $(S \geq \tau_n)$, $S \wedge \tau_{n+1} = (\tau_n + R_n) \wedge \tau_{n+1}$.*

Proof. From Lemma N. , because S is $\mathcal{F}_S^N$ measurable, it is $\sigma(N_{t\wedge S}, t \geq 0)$ measurable. From Lemma B. and Lemma M. there exists a measurable function ϕ such that

$$S = \phi(N_{t\wedge S}, t \in I),$$

for some countable set I. If we consider S on $\{\tau_n \leq S < \tau_{n+1}\}$ then

$$SI_{\{\tau_n \leq S < \tau_{n+1}\}} = \psi_n(\tau_0, \ldots, \tau_n) I_{\{\tau_n \leq S < \tau_{n+1}\}},$$

for some measurable ψ_n. If we take $R_n = (\psi_n - \tau_n)_+$ then on $\{S \geq \tau_n\}$ the above equality gives

$$S \wedge \tau_{n+1} = (\tau_n + R_n) \wedge \tau_{n+1}.$$

□

Proof of Theorem K. .

We use Theorem J. Lemma L. to see that it is enough to show that

$$E(N_{S\wedge\tau_n}) = E(A_{S\wedge\tau_n}), n \geq 1,$$

for arbitrary $(\mathcal{F}_t^N)$ stopping times S, where $A_t = \int_0^t \lambda_u$ du.
Let S be an arbitrary $(\mathcal{F}_t^N)$ stopping time , and $\{R_n\}$ the sequence from Lemma O. . We first compute $E(N_{S\wedge\tau_n})$:

$$E(N_{S\wedge\tau_n}) = E(\sum_{j=0}^{n-1}(N_{S\wedge\tau_{j+1}} - N_{S\wedge\tau_j})I_{(S\geq\tau_j)}) =$$

$$= E(\sum_{j=0}^{n-1} I_{(R_j\geq\tau_{j+1}-\tau_j)}I_{(S\geq\tau_j)}) = E(\sum_{j=0}^{n-1} P(I_{(R_j\geq\tau_{j+1}-\tau_j)}I_{(S\geq\tau_j)} \mid \mathcal{F}_{\tau_j}^N)) =$$

$$= E(\sum_{j=0}^{n-1}\int_0^{R_j} f^{(j+1)}(u)du I_{(S\geq\tau_j)}),$$

from the definition of $f^{(n+1)}$, and $\mathcal{F}_{\tau_j}^N$ measurability of $I_{(S\geq\tau_j)}$. To compute $E(A_{S\wedge\tau_n})$, let $r_{j+1}(u) = f^{(j+1)}(u)/(\int_u^\infty f^{(j+1)}(x)dx), j \geq 1$. From our assumptions we have

$$E(A_{S\wedge\tau_n}) = E(\sum_{j=0}^{n-1}\int_0^{R_j\wedge(\tau_{j+1}-\tau_j)} r_{j+1}(u)du I_{(S\geq\tau_j)}) =$$

$$= E(\sum_{j=0}^{n-1} E(\int_0^{R_j\wedge(\tau_{j+1}-\tau_j)} r_{j+1}(u)du I_{(S\geq\tau_j)} \mid \mathcal{F}_{\tau_j}^N)) =$$

$$= E(\sum_{j=0}^{n-1} I_{(S\geq\tau_j)}\int_0^\infty\int_0^{R_j\wedge x} r_{j+1}(u)du f^{(j+1)}(x)dx) =$$

$$= E(\sum_{j=0}^{n-1} I_{(S\geq\tau_j)}\int_0^{R_j\wedge x}\int_0^\infty I_{(0,x)}(u) f^{(j+1)}(x)dx r_{j+1}(u)du) =$$

$$= E(\sum_{j=0}^{n-1} \int_0^{R_j} f^{(j+1)}(u) du I_{(S \geq \tau_j)}),$$

where we used Fubini's theorem, $\mathcal{F}_{\tau_j}^N$ measurability of $I_{(S \geq \tau_j)}$, and the definition of r_j's. □

From Theorem K. we have the following corollary

Corollary P. *Under the assumptions of Theorem K., if a counting process* (N_t) *has* $(\mathcal{F}_t^N)$ *stochastic intensity* (λ_t) *then its compensator* (A_t) *can be written in the form*

$$A(t) = \sum_{n \geq 0} a_n(t; \tau_0, \ldots, \tau_n) I_{(\tau_n \leq t < \tau_{n+1})}, \tag{6}$$

where $\{a_n, n \geq 0\}$ *is a family of multidimensional measurable functions*

A similar concept to transition function family for Markov chains is a **compensator function family.**

We say that that the family of functions

$$\{a_n(t; t_0, \ldots, t_n), 0 \leq t_0 < t_1 < \ldots < t_n, n \geq 0, t \geq 0\}$$

is a **compensator function family** of a counting process (N_t) if

- $a_n(t; t_0, \ldots, t_n)$ is nondecreasing and right continuous for each fixed $t_1, \ldots, t_n$;
- $a_{n+1}(t; t_0, \ldots, t_k, s, t_{k+1}, \ldots, t_n) = a_k(t; t_0, \ldots, t_k)$, for $t \leq s$;
- for each $n \geq 0$, $a_n(t; t_1, \ldots, t_n)$ is measurable function of $(t, t_1, \ldots, t_n)$;
- $\sum_{n \geq 0} a_n(t; \tau_0, \ldots, \tau_n) I_{(\tau_n < t \leq \tau_{n+1})}$ is a version of $(\mathcal{F}_t^N)$ compensator of (N_t).

We shall see later that compensator function families are useful in stochastic ordering of point processes, similarly as transition function families are useful in stochastic ordering of Markov chains.

Examples.

1. Let $\{X_n\}_{n \geq 1}$ be a sequence of i.i.d. positive random variables. Define $\tau_n = \sum_{i=1}^n X_i, n \geq 1$, and $\tau_0 = 0$. The corresponding counting process (N_t) is called a non-delayed renewal process. Suppose that the distribution of X_1 has a positive density f, i.e. $F(x) = \int_0^x f(u) du$, where F is its distribution function. Denote by r the corresponding failure rate. Under these assumptions (N_t) has an $(\mathcal{F}_t^N)$ stochastic intensity (λ_t) given by

$$\lambda_t = \sum_{i=0}^{\infty} r(t - \tau_i) I_{(\tau_i \leq t < \tau_{i+1})}. \tag{7}$$

Recall that for a renewal process we define the age process by $\mathcal{A}(t) = t - \sum_{i=0}^{N_t} X_i$. Using this notation we have

$$\lambda_t = r(\mathcal{A}(t)), t \geq 0.$$

2. Let $(X_t, t \geq 0)$ be a continuous time Markov chain with a uniform intensity matrix $\mathbf{Q}$. By uniformization we can have a version of (X_t) with right continuous trajectories. For such a version of (X_t) we define a counting process

$$N_t^f = \sum_{u \leq t} f(X_{u-}, X_u),$$

where f is a zero-one valued function such that $f(i,i) = 0, i \in$E. This process counts certain, selected by f , transitions of (X_t). From the definition, (N_t^f) is adapted to $(\mathcal{F}_t^X)$. By the formula (2) , (N_t^f) has an $(\mathcal{F}_t^X)$ stochastic intensity (λ_t) given by

$$\lambda_t = \sum_{j \neq X_t} q_{X_t j} f(X_t, j). \tag{8}$$

If $\alpha(i) = \sum_{j \neq i} q_{ij} f(i,j), i \in E$, then we can rewrite this intensity as

$$\lambda_t = \alpha(X_t).$$

For a detailed treatment also in many dimensions see Last and Brandt (1994).

2.6 Markovian queues and Jackson networks

Basic models

In this paragraph we shall recall intensity matrices, which serve for the most common continuous time Markov chains encountered in applied probability.

1. (Poisson process)
Consider $E = \{0, 1, 2, \ldots\}$ and the following intensity matrix

$$\mathbf{Q} = \begin{bmatrix} -\lambda & \lambda & 0 & 0 & \cdots \\ 0 & -\lambda & \lambda & 0 & \cdots \\ \vdots & \ddots & \ddots & \ddots & \ddots \end{bmatrix}, \tag{1}$$

where $\lambda > 0$. This is of course a uniform intensity matrix. The corresponding (see Theorem 2.4.F.) backward differential equation $\frac{d\mathbf{P}_t}{dt} = \mathbf{QP}_t$ takes the following form

$$\frac{dp_{ij}(t)}{dt} = -\lambda p_{ij}(t) + \lambda p_{i+1,j}(t), \; i,j \in E, \; t \geq 0,$$

with the initial condition $\mathbf{P}_0 = \mathbf{I}$.
From this we have

$$\frac{dp_{00}(t)}{dt} = -\lambda p_{00}(t),$$

with $p_{00}(0) = 1$, which implies

$$p_{00}(t) = e^{-\lambda t}, \; t \geq 0,$$

and inductively

$$p_{0j}(t) = \frac{(\lambda t)^j}{j!} e^{-\lambda t}, \; t \geq 0.$$

The above equality says that if a Poisson process starts with 0 (at time 0), then its distribution of the number of points in the interval $[0,t]$ is Poisson.

2. (birth process)
Consider $E = \{0, 1, 2, \ldots\}$ and

$$\mathbf{Q} = \begin{bmatrix} -\lambda_0 & \lambda_0 & 0 & 0 & \cdots \\ 0 & -\lambda_1 & \lambda_1 & 0 & \cdots \\ \vdots & \ddots & \ddots & \ddots & \ddots \end{bmatrix},$$

for $\lambda_i > 0,\ i \in E$, and $\sup_i \lambda_i < \infty$. The corresponding backward equations are

$$\frac{dp_{ij}(t)}{dt} = -\lambda_i p_{ij}(t) + \lambda_i p_{i+1,j}(t),\ i,j \in E,\ t \geq 0,$$

with the initial condition $\mathbf{P}_0 = \mathbf{I}$.
It is known (see e.g. Feller (1968)) that $\sum_j p_{0j}(t) = 1$ if and only if $\sum_j \frac{1}{\lambda_j} = \infty$. In terms of the corresponding Markov chain, which starts from 0 at the origin, it means that the process does not "explode" in finite time t if and only if $\sum_j \frac{1}{\lambda_j} = \infty$. Of course this is the case if $\sup_i \lambda_i < \infty$.

3. (birth and death process)
Consider $E = \{0, 1, 2, \ldots\}$ and

$$\mathbf{Q} = \begin{bmatrix} -\lambda_0 & \lambda_0 & 0 & 0 & \cdots & \cdots \\ \mu_1 & -(\lambda_1 + \mu_1) & \lambda_1 & 0 & \cdots & \cdots \\ 0 & \mu_2 & -(\lambda_2 + \mu_2) & \lambda_2 & 0 & \cdots \\ \vdots & \ddots & \ddots & \ddots & \ddots & \ddots \end{bmatrix},$$

for $\lambda_i, \mu_i > 0,\ i \in E$ and $\sup_i(\lambda_i + \mu_i) < \infty$. The corresponding forward equations are

$$\frac{dp_{ij}(t)}{dt} = \lambda_{j-1} p_{i,j-1}(t) + \mu_{j+1} p_{i,j+1} - (\lambda_j + \mu_j) p_{ij}(t),\ i,j \in E,\ j \geq 1,\ t \geq 0,$$

and

$$\frac{dp_{i0}(t)}{dt} = \mu_1 p_{i,1} - \lambda_0 p_{i0}(t),\ i \in E,\ t \geq 0,$$

with the initial condition $\mathbf{P}_0 = \mathbf{I}$.
Under the assumption $\sup_i(\lambda_i + \mu_i) < \infty$ there exists a unique solution of these equations such that $\sum_j p_{ij}(t) = 1$, and the limits $\lim_{t\to\infty} p_{ij}(t) = \pi_j$ (see e.g. Feller (1968)). Therefore, for π_j's we have

$$\lambda_0 \pi_0 = \mu_1 \pi_1,$$

$$(\lambda_j + \mu_j)\pi_j = \lambda_{j-1}\pi_{j-1} + \mu_{j+1}\pi_{j+1},\ j \geq 1.$$

This leads to

$$\pi_1 = \frac{\lambda_0}{\mu_1}\pi_0,$$

and by induction

$$\pi_n = \pi_0 \frac{\lambda_0 \cdots \lambda_{n-1}}{\mu_1 \cdots \mu_n},\ n \geq 1.$$

This will be a stationary distribution if $\sum_{n\geq 0}\pi_n = 1$, which implies that

$$\pi_0 = \frac{1}{1+\sum_{n\geq 1}\frac{\lambda_0\cdots\lambda_{n-1}}{\mu_1\cdots\mu_n}},$$

should be positive and finite, thus we should have

$$\sum_{n\geq 1}\frac{\lambda_0\cdots\lambda_{n-1}}{\mu_1\cdots\mu_n} < \infty.$$

Markovian queues

4. ($M/M/s$ FCFS queue)
Consider $E = \{0,1,2,\ldots\}$, and

$$\mathbf{Q} = \begin{bmatrix} -\lambda & \lambda & 0 & 0 & \cdots & \cdots \\ \mu_1 & -(\lambda+\mu_1) & \lambda & 0 & \cdots & \cdots \\ 0 & \mu_2 & -(\lambda+\mu_2) & \lambda & 0 & \cdots \\ \vdots & \ddots & \ddots & \ddots & \ddots & \ddots \end{bmatrix},$$

where $\mu_i = (i\wedge s)\mu,\ i\geq 1$, for some $\mu > 0$, and $s\geq 1,\ \lambda > 0$. This is a special case of a birth-death type matrix, which is uniform. The corresponding forward equations are of the following form (cf. example 3.)

$$\frac{dp_{ij}(t)}{dt} = \lambda p_{i,j-1}(t) + \mu_{j+1}p_{i,j+1} - (\lambda+\mu_j)p_{ij}(t),\ i,j\in E,\ j\geq 1,\ t\geq 0,$$

and

$$\frac{dp_{i0}(t)}{dt} = \mu p_{i,1} - \lambda p_{i0}(t),\ i\in E,\ t\geq 0,$$

with the initial condition $\mathbf{P}_0 = \mathbf{I}$.
The solution for $s > 1$ is rather complicated (see e.g. Saaty (1960)). However it is easy to find the limits $\lim_{t\to\infty} p_{ij}(t) = \pi_j$ (which do exist in this case). We have from the forward equations (letting $t\to\infty$)

$$\lambda\pi_0 = \mu\pi_1,$$

$$(\lambda + j\mu)\pi_j = \lambda\pi_{j-1} + (j+1)\mu\pi_{j+1},$$

for $j\leq s$, and

$$(\lambda + s\mu)\pi_j = \lambda\pi_{j-1} + s\mu\pi_{j+1},$$

for $j\geq s$. This results in the following solution

$$\pi_j = \pi_0\frac{(\frac{\lambda}{\mu})^j}{j!},$$

for $j\leq s$, and

$$\pi_j = \pi_0\frac{(\frac{\lambda}{\mu})^j}{s!s^{j-s}},$$

for $j \geq s$. The normalization $\sum_{j\geq 0}\pi_j = 1$ implies

$$\pi_0 = (\sum_{j=0}^{s-1}\frac{1}{j!}(\frac{\lambda}{\mu})^j + \frac{s^s}{s!}\sum_{j=s}^{\infty}(\frac{\lambda}{\mu s})^j)^{-1},$$

which is positive and finite if $\frac{\lambda}{\mu} < s$.

5. (loss system $M/M/1/m$)

Consider a birth and death process with the state space $E = \{0, \dots, m\}$, and the following intensity matrix

$$\mathbf{Q} = \begin{bmatrix} -m\lambda & m\lambda & 0 & 0 & \cdots & \cdots \\ \mu & -((m-1)\lambda+\mu) & (m-1)\lambda & 0 & \cdots & \cdots \\ 0 & \mu & -((m-2)\lambda+\mu) & (m-2)\lambda & 0 & \cdots \\ \vdots & \ddots & \ddots & \ddots & \ddots & \ddots \\ 0 & \cdots & \cdots & \cdots & \mu & -\mu \end{bmatrix},$$

where $\mu > 0$, $\lambda > 0$. The corresponding forward equations are of the following form

$$\frac{dp_{ij}(t)}{dt} = (m-j+1)\lambda p_{i,j-1}(t) + \mu p_{i,j+1} - ((m-j)\lambda+\mu)p_{ij}(t),\ i,j \in E,\ 1 \leq j \leq m-1,\ t \geq 0,$$

$$\frac{dp_{i0}(t)}{dt} = \mu p_{i,1} - m\lambda p_{i0}(t),\ i \in E,\ t \geq 0,$$

and

$$\frac{dp_{im}(t)}{dt} = \lambda p_{i,m-1}(t) - \mu p_{im}(t),\ i \in E,\ t \geq 0,$$

with the initial condition $\mathbf{P}_0 = \mathbf{I}$.
For the limits $\lim_{t\to\infty} p_{ij}(t) = \pi_j$ (which do exist in this case because the state space E is finite) we have

$$m\lambda\pi_0 = \mu\pi_1,$$
$$((m-j)\lambda+\mu)\pi_j = (m-j+1)\lambda\pi_{j-1} + \mu\pi_{j+1},\ 1 \leq j \leq m-1,$$
$$\mu\pi_m = \lambda\pi_{m-1}.$$

This implies

$$\pi_{m-j} = \frac{1}{j!}(\frac{\mu}{\lambda})^j\pi_m,\ 1 \leq j \leq m,$$

$$\pi_m = (1 + \frac{\mu}{\lambda} + \ldots + \frac{1}{m!}(\frac{\mu}{\lambda})^m)^{-1},$$

which is Erlang's loss formula for this system.

6. ($M/M/\infty$ queue)

Consider a birth and death process with the state space $E = \{0, 1, \ldots\}$, and the following nonuniform intensity matrix

$$\mathbf{Q} = \begin{bmatrix} -\lambda & \lambda & 0 & 0 & \cdots & \cdots \\ \mu & -(\lambda+\mu) & \lambda & 0 & \cdots & \cdots \\ 0 & 2\mu & -(\lambda+2\mu) & \lambda & 0 & \cdots \\ \vdots & \ddots & \ddots & \ddots & \ddots & \ddots \end{bmatrix},$$

where $\mu > 0,\ \lambda > 0$. The corresponding forward equations are of the following form

$$\frac{dp_{ij}(t)}{dt} = \lambda p_{i,j-1}(t) + (j+1)\mu p_{i,j+1} - (\lambda + j\mu)p_{ij}(t),\ i,j \in E,\ j \geq 1,\ t \geq 0,$$

$$\frac{dp_{i0}(t)}{dt} = \mu p_{i,1} - \lambda p_{i0}(t),\ i \in E,\ t \geq 0,$$

with the initial condition $\mathbf{P}_0 = \mathbf{I}$.
For the limits $\lim_{t\to\infty} p_{ij}(t) = \pi_j$ (which do exist in this case, see e.g. Prabhu (1965)) we have

$$\lambda\pi_0 = \mu\pi_1,$$

$$(\lambda + j\mu)\pi_j = \lambda\pi_{j-1} + (j+1)\mu\pi_{j+1},\ j \geq 1.$$

This implies

$$\pi_j = \frac{1}{j!}(\frac{\lambda}{\mu})^j\pi_0,\ j \geq 1,$$

hence

$$\pi_j = e^{-(\lambda/\mu)}\frac{1}{j!}(\frac{\lambda}{\mu})^j,$$

that is, the stationary distribution for this queue is Poisson.

Jackson networks

Consider $E = Z^n$, and the following intensity matrix $\mathbf{Q} = [q_{\mathbf{ij}}]_{\mathbf{i},\mathbf{j}\in E}$:

$$q_{\mathbf{ij}} \doteq \begin{cases} p_{0k}\lambda & \mathbf{j} = \mathbf{i} + \mathbf{e}_k, \\ p_{k0}\mu_k & \mathbf{j} = \mathbf{i} - \mathbf{e}_k,\ i_k \geq 1, \\ p_{kl}\mu_k & \mathbf{j} = \mathbf{i} + \mathbf{m}_{kl},\ i_k \geq 1, \\ 0 & elsewhere, \end{cases}$$

where $\mathbf{i} = (i_1, \ldots, i_n) \in E,\ \mathbf{i} \neq \mathbf{j},\ \mathbf{e}_k$ is the vector with 1 in position k and 0's elsewhere, $\mathbf{m}_{kl} = \mathbf{e}_l - \mathbf{e}_k,\ \lambda, \mu > 0$, and $\mathbf{P} = [p_{kl}]_{0\leq k,l\leq n}$ is an irreducible probability matrix with $p_{00} = 0$. For $\mathbf{i} = \mathbf{j}$ we take

$$q_{\mathbf{ii}} = -\sum_{\mathbf{j}\neq\mathbf{i}} q_{\mathbf{ij}},\ \mathbf{i} \in E.$$

Thus $\mathbf{Q}$ is regular and uniform. The corresponding (see Corollary G.) continuous time Markov chain $(\mathbf{X}(t) = (X_1(t), \ldots, X_n(t)),\ t \geq 0)$ is called **Jackson network**; this process corresponds to a network of n nodes representing service stations. Customer enter the nodes $1, \ldots, n$ from outside the network according to independent Poisson processes with the respective rates $p_{01}\lambda, \ldots, p_{0n}\lambda$, some of which may be zero. Each node k operates as an isolated single server whose service times are independent exponential random variables with mean $(\mu_k)^{-1}$. Customers are served one at a time, under any priority scheme. A customer after being served at node k, goes immediately to node l with probability $p_{kl},\ l = 1, \ldots, n$, or exits the network with probability p_{k0}. $\mathbf{X}(t)$ denotes the numbers of customers at the respective nodes at time t.

2.7 Poissonian flows and product formula

Let $(X_t, t \geq 0)$ be a continuous time Markov chain with a uniform intensity matrix $\mathbf{Q}$. By uniformization we can have a version of (X_t) with right continuous trajectories. For such a version of (X_t) we define a counting process

$$N_t^f = \sum_{u \leq t} f(X_{u-}, X_u),$$

where f is a zero-one valued function such that $f(i,i) = 0$, $i \in$E. This process counts certain, selected by f , transitions of (X_t). From the definition, (N_t^f) is adapted to $(\mathcal{F}_t^X)$. By the formula 2.5. (8) , (N_t^f) has an $(\mathcal{F}_t^X)$ stochastic intensity (λ_t) given by

$$\lambda_t = \sum_{j \neq X_t} q_{X_t j} f(X_t, j). \tag{1}$$

If $\alpha(i) = \sum_{j \neq i} q_{ij} f(i,j), i \in E$, then we can rewrite this intensity as

$$\lambda_t = \alpha(X_t). \tag{2}$$

We turn our attention now to the question : when (N_t^f) is a stationary Poisson process, i.e. when it is $(\mathcal{F}_t^{N^f})$ Poisson ? Note that if (N_t^f) is $(\mathcal{F}_t^X)$ Poisson then it is $(\mathcal{F}_t^{N^f})$ Poisson, (see Section 2.5. after Theorem G.), however many processes of interest fail to be $(\mathcal{F}_t^X)$ Poisson being though $(\mathcal{F}_t^{N^f})$ Poisson. Therefore another approach is needed. An interesting insight to this problem we can gain by reversing processes in time. Consider now the process which is a continuous time Markov chain with the intensity matrix $\mathbf{Q}'$, given in Theorem 2.4.L., which starts with $\boldsymbol{\pi}$, i.e. we consider a process $(X_t^{\leftarrow})$, which is the time reverse process of (X_t). We can assume that $(X_t^{\leftarrow})$ has also right continuous trajectories. For $(X_t^{\leftarrow})$, define

$$N_t^{f\leftarrow} = \sum_{u \leq t} f^{\leftarrow}(X_{u-}^{\leftarrow}, X_u^{\leftarrow}),$$

where $f^{\leftarrow}(i,j) = f(j,i)$. Then it is clear that $(N_t^{f\leftarrow})$ counts "the same " transitions as (N_t^f), and therefore $(N_t^{f\leftarrow})$ is standard Poisson if and only if (N_t^f) is standard Poisson. The usefulness of $(N_t^{f\leftarrow})$ is that we sometimes can easily show that it is $(\mathcal{F}_t^{X\leftarrow})$ Poisson, which implies that it is stationary Poisson and from this we have that (N_t^f), our process of original interest, is stationary Poisson.

Following Serfozo (1989) we define some useful concepts of independence for histories. For (N_t^f), let $\mathcal{F}'^{N^f}_t = \sigma(N_s^f, s \geq t)$, we say that the future of (N_t^f) is independent of the past of (X_t) (in symbols $N_+^f \perp X_-$) if $\mathcal{F}'^{N^f}_t$ is independent of $\mathcal{F}_t^X$, for all $t \geq 0$. Similarly we define $N_-^f \perp X_+$. The following theorem is proved in Serfozo (1989)

Theorem A.

(i) *If $N_+^f \perp X_-$ or $N_-^f \perp X_+$ then (N_t^f) is $(\mathcal{F}_t^{N^f})$ Poisson process;*

(ii) *If (N_t^f) is $(\mathcal{F}_t^X)$ Poisson process then $N_+^f \perp X_-$;*

(iii) *If* (N_t^{f-}) *is* $(\mathcal{F}_t^{X-})$ *Poisson then* $N_-^f \perp X_+$.

From the above theorem, Theorem 2.5.I. and (1) we have the following corollary.

Corollary B. *If the function* $\alpha^{\leftarrow}(i) = \sum_{j\neq i} q'(i,j) f^{\leftarrow}(i,j)$ *is constant then* (N_t^{f-}) *and* (N_t^f) *are stationary Poisson processes.*

The theory of stochastic intensities can be generalized to multidimensional contexts. In many models, it is natural to associate with each jump of a point process a "mark", which is usually a random vector. In such a situation we consider a family of point processes with marks in specified sets. We can describe this formally as follows. Let $(\Omega, \mathcal{F}, P)$ be a probability space, and $(\mathcal{F}_t, t \geq 0)$ a history on this space. Suppose that $(\tau_n)_{n\geq 1}$ is a sequence of $(\mathcal{F}_t)$ stopping times, such that $\tau_1 < \tau_2 < \ldots, \tau_n \to \infty, n \to \infty$ $(\tau_0 = 0)$. Denote by $\{Z_n\}_{n\geq 1}$ an arbitrary sequence of random variables with values in a measurable mark space $(S, \mathcal{S})$. For each $B \in \mathcal{S}$, define

$$N_t(B) = \sum_{n\geq 1} I_{(\tau_n \leq t)} I_{(Z_n \in B)},$$

and

$$\mathcal{F}_t^{\mathrm{N}} = \sigma(N_s(B), s \leq t, B \in \mathcal{S})$$

A complete description of this multidimensional situation we obtain by introducing the following counting measure on $\mathcal{B}^1 \otimes \mathcal{S}$

$$\mathbf{N}(A \times B) = \sum_{n\geq 1} I_{(\tau_n \in A)} I_{(Z_n \in B)},$$

for $A \in \mathcal{B}^1, B \in \mathcal{S}$. We call $\mathbf{N}$ a **marked point process.**
Suppose that $\mathcal{F}_t^{\mathrm{N}} \subseteq \mathcal{F}_t$ for some history $(\mathcal{F}_t)$ on $(\Omega, \mathcal{F}, P)$. A multidimensional analog of stochastic intensity is the following.

Suppose that for all $B \in \mathcal{S}, N_t(B)$ has $(\mathcal{F}_t)$ predictable intensity $(\lambda_t(B))$, such that for all fixed ω, t, $\lambda_t(.)$ is a measure on $\mathcal{S}$. We say that $\mathbf{N}$ admits $(\mathcal{F}_t)$ **intensity kernel** $(\lambda_t(ds))$.
From a multidimensional version of Theorem 2.5.I. we have a useful characterization of Poisson processes (see e.g. Bremaud (1981), VIII, E3).

Theorem C. *If a marked point process* $\mathbf{N}$ *admits* $(\mathcal{F}_t)$ *intensity kernel* $(\lambda_t(ds))$ *of the form*

$$\lambda_t(ds) = \lambda F(ds),$$

where $\lambda > 0$, *and* F *is a probability measure on* $(S, \mathcal{S})$ *then for any collection* $B_1, \ldots, B_k$ *of disjoint sets,* $N_t(B_1), \ldots, N_t(B_k)$ *are independent* $(\mathcal{F}_t^{\mathrm{N}})$ *Poisson processes.*

An important special case is a marked point process describing selected jumps in a Markov chain. Let (X_t) be a continuous time Markov chain with an intensity matrix $\mathbf{Q}$, which is uniform. Define

$$N_t^{f,h}(B) = \sum_{u\leq t} f(X_{u-}, X_u) I_{(h(X_{u-}, X_u) \in B)},$$

where f is zero-one valued function such that $f(i,i) = 0, i \in E$, and h is a function with values in $(S, \mathcal{S})$. This is a process counting selected jumps of the process (X_t) (selected by f) with marks assigned at jump points to be in a specified set B (marks are assigned by h). Many processes of interest can be modeled this way.
In the case when (X_t) is stationary, the following functions will be useful.

$$\alpha(i,B) = \sum_{j \neq i} q_{i,j} f(i,j) I_{(h(i,j) \in B)},$$

$$\alpha^{\leftarrow}(i,B) = \sum_{j \neq i} q'_{ij} f^{\leftarrow}(i,j) I_{(h(j,i) \in B)}, \tag{3}$$

where $\mathbf{Q}'$ is the intensity matrix of the time reverse process of (X_t).
Analogously to the formula (2) we have

Lemma D. *The marked point process $\mathbf{N}^{f,h}$ admits $(\mathcal{F}_t^X)$ intensity kernel $(\lambda_t(ds))$ given by*

$$\lambda_t(B) = \alpha(X_t, B), \;\; B \in \mathcal{S}.$$

A multidimensional analogy of Corollary B. is as follows.

Corollary E. *Suppose (X_t) is stationary. If*

$$\alpha(i,B) = \lambda F(B),$$

or

$$\alpha^{\leftarrow}(i,B) = \lambda F(B),$$

for $B \in \mathcal{S}$, where $\lambda > 0$, and F is a probability measure then $(N_t(B_1)), \ldots, (N_t(B_k))$ are independent $(\mathcal{F}_t^{N^{f,h}})$ Poisson processes, for all disjoint $B_1, \ldots, B_k \in S$, $k \geq 1$.

We shall apply the above corollary to Jackson networks. Before this we shall prove the celebrated product formula of Jackson networks.

Consider again $E = Z^n$, and a Markov process $\mathbf{X}(t)$ with the following intensity matrix $\mathbf{Q} = [q_{\mathbf{ij}}]_{\mathbf{i},\mathbf{j} \in E}$:

$$q_{\mathbf{ij}} = \begin{cases} p_{0k}\lambda & \mathbf{j} = \mathbf{i} + \mathbf{e}_k, \\ p_{k0}\mu_k & \mathbf{j} = \mathbf{i} - \mathbf{e}_k, \; i_k \geq 1, \\ p_{kl}\mu_k & \mathbf{j} = \mathbf{i} + \mathbf{m}_{kl}, \; i_k \geq 1, \\ 0 & elsewhere, \end{cases}$$

where $\mathbf{i} = (i_1, \ldots, i_n) \in E$, $\mathbf{i} \neq \mathbf{j}$, $\mathbf{e}_k$ is the vector with 1 in position k and 0's elsewhere, $\mathbf{m}_{kl} = \mathbf{e}_l - \mathbf{e}_k$, $\lambda, \mu > 0$, and $\mathbf{P} = [p_{kl}]_{0 \leq k,l \leq n}$ is an irreducible probability matrix with $p_{00} = 0$. For $\mathbf{i} = \mathbf{j}$ we take

$$q_{\mathbf{ii}} = -\sum_{\mathbf{j} \neq \mathbf{i}} q_{\mathbf{ij}}, \; \mathbf{i} \in E.$$

Consider also the following equations

$$\alpha_j = \lambda p_{0j} + \sum_{k=1}^{n} \alpha_k p_{kj}, \; j = 0, \ldots, n. \tag{4}$$

We call (4) **the traffic equations**. These equations have a unique solution. Indeed, let $\beta = (\beta_0, \ldots, \beta_n)$ be a probability distribution satisfying $\beta = \beta\mathbf{P}$ (which does exist because we assumed that $\mathbf{P}$ is an irreducible probability matrix) then $\alpha_j = \frac{\lambda}{\beta_0}\beta_j$ is a unique solution for (4).
Let $\rho_k = \alpha_k/\mu_k$, $k = 1, \ldots, n$ be the corresponding **traffic intensities** through the respective nodes. The following theorem gives a **product formula** for the invariant distribution with respect to $\mathbf{Q}$.

Theorem F. (Product formula) *The Jackson network process $(\mathbf{X}(t))$ has the stationary distribution given by*

$$\pi_{\mathbf{i}} = \prod_{k=1}^{n}(1-\rho_k)(\rho_k)^{i_k}, \ \mathbf{i} \in E.$$

Proof. We apply Corollary 2.4.M. to prove this theorem. Applying the formula 2.4. (3) to the above $\pi'_{\mathbf{i}}$s we find that the corresponding $\mathbf{Q}'$ matrix is given by

$$q'_{\mathbf{ij}} = \begin{cases} (\rho_k)^{-1}p_{0k}\lambda & \mathbf{j} = \mathbf{i} - \mathbf{e}_k, \ i_k \geq 1, \\ \rho_k p_{k0}\mu_k & \mathbf{j} = \mathbf{i} + \mathbf{e}_k, \\ \rho_k(\rho_l)^{-1}p_{kl}\mu_k & \mathbf{j} = \mathbf{i} + \mathbf{m}_{lk}, \ i_l \geq 1, \\ 0 & elsewhere. \end{cases} \tag{5}$$

We have to prove that $\mathbf{Q}'$ is an intensity matrix, which is equivalent here to check whether $-\sum_{\mathbf{j}\neq\mathbf{i}} q'_{\mathbf{ij}} = q_{\mathbf{ii}} = q'_{\mathbf{ii}}$.
We start with computing $q'_{\mathbf{ii}}$.

$$\begin{aligned} q'_{\mathbf{ii}} &= \sum_{k\geq 1}[((\rho_k)^{-1}p_{0k}\lambda + \rho_k p_{k0}\mu_k) + \sum_{l\geq 1}\rho_k(\rho_l)^{-1}p_{kl}\mu_k] = \\ &= \sum_{k\geq 1}p_{k0}\alpha_k + \sum_{l\geq 1}[(\rho_l)^{-1}(p_{0l}\lambda + \sum_{k\geq 1}\alpha_k(\rho_l)^{-1}p_{kl})] = \\ &= \alpha_0 + \sum_{l\geq 1}(\rho_l)^{-1}\alpha_l = \alpha_0 + \sum_{l\geq 1}\mu_l. \end{aligned}$$

On the other hand we have

$$\begin{aligned} q_{\mathbf{ii}} &= \sum_{k\geq 1}p_{0k}\lambda + \sum_{k\geq 1}\sum_{l\geq 0}p_{kl}\mu_k = \\ &= \sum_{k\geq 1}(\alpha_k - \sum_{j\geq 1}\alpha_j p_{jk}) + \sum_{k\geq 1}\mu_k = \\ &= \sum_{k\geq 1}\alpha_k - \sum_{j\geq 1}\alpha_j(1-p_{j0}) + \sum_{k\geq 1}\mu_k = \\ &= \sum_{j\geq 1}\alpha_j p_{j0} + \sum_{k\geq 1}\mu_k = \alpha_0 + \sum_{k\geq 1}\mu_k, \end{aligned}$$

which completes the proof. □

Now, by an appropriate specification of f and h in $N_t^{f,h}$, consider the departure processes from a Jackson network, which are given by

$$N_{j0}(t) = \sum_{s\leq t} I_{(\mathbf{X}(s)=\mathbf{X}(s-)-\mathbf{e}_j)},$$

where $\mathbf{e}_j$ denotes the jth unit vector of E. As h we take here $h(\mathbf{i},\mathbf{j}) = \mathbf{i} - \mathbf{j}$, and f indicates all jumps of $\mathbf{X}$ $(B = \{\mathbf{e}_j\})$.

Theorem G. (Poissonian flows) *The vector of departure processes from a stationary Jackson network is a vector of independent Poisson processes.*

Proof. From (3) and (5) we immediately have for our f, h

$$\alpha^{\leftarrow}(\mathbf{i}, \{\mathbf{e}_j\}) = \alpha_j p_{j0}.$$

Now from Corollary E. we conclude that $(N_{10}, \ldots, N_{n0})$ is a vector of independent Poisson processes. □

PROBLEMS AND REMARKS

A. Consider a Jackson network $\mathbf{X}(t) = (X_1(t), X_2(t))$ with the following intensity matrix $\mathbf{Q} = [q_{\mathbf{ij}}]_{\mathbf{i},\mathbf{j} \in Z^2}$:

$$q_{\mathbf{ij}} = \begin{cases} p_{0k} & \mathbf{j} = \mathbf{i} + \mathbf{e}_k, \\ p_{k0}\mu_k & \mathbf{j} = \mathbf{i} - \mathbf{e}_k,\ i_k \geq 1, \\ p_{kl}\mu_k & \mathbf{j} = \mathbf{i} + \mathbf{m}_{kl},\ i_k \geq 1, \\ 0 & elsewhere, \end{cases}$$

where $\mathbf{i} \neq \mathbf{j}$, $\mathbf{e}_k$ is the vector with 1 in position k and 0's elsewhere, $\mathbf{m}_{kl} = \mathbf{e}_l - \mathbf{e}_k$, $\mu_1 = 1$, $\mu_2 = 2$ and $\mathbf{P} = [p_{kl}]_{0 \leq k,l \leq 2}$ is given by

$$\begin{bmatrix} 0 & \frac{1}{2} & \frac{1}{2} \\ 0 & 0 & 1 \\ 1 & 0 & 0 \end{bmatrix}$$

Find stationary distribution $\mathbf{Pr}(X_1(t) = k, X_2(t) = l)$.

B. Let $(X(t))$ be a Markov process with transition rates

$$\begin{aligned} q_{n,n+1} &= \lambda \quad n \in Z_+ \\ q_{n,n-K} &= \mu \quad n \geq K \\ q_{n,0} &= \mu \quad n < K \end{aligned}$$

where λ, μ, K are positive. This process represents the number of customers in an $M/M/1$ batch service queue. The point process N of times at which batches of size K depart from the system is Poisson. [Serfozo (1989)]. (For $K = 1$ see Burke (1956), Reich (1957)).

2.8 Stochastic ordering of Markov processes

State spaces

(E, M) is a generic symbol for a measurable state space. This space is usually equipped with an order relation $\prec$, which is

- preorder i.e.

$$x \prec x, (x \prec y, y \prec z) \to x \prec z, \text{ or}$$

- partial order i.e. preorder and

$$(x \prec y, y \prec x) \to x = y, \text{ or}$$

- total order i.e. order and $x \prec y$ or $y \prec x$.

If E is a topological space then M denotes Borel σ field. Additionally one considers concepts relating order and topology (Nachbin (1965)).

1. $\prec$ is **closed** if $\{x \prec y\}$ is closed in E^2.
2. $(E, \prec)$ is **compact ordered** if E is compact and $\prec$ is closed.
3. $(E, \prec)$ is **normally ordered** if for each pair of disjoint closed sets $\mathbf{F}$ and $\mathbf{G}$ there exist open sets $\mathbf{U}_F$ and $\mathbf{U}_G$ disjoint and including respectively the above sets , where $\mathbf{F}$ and $\mathbf{U}_F$ are decreasing, $\mathbf{G}$ and $\mathbf{U}_G$ are increasing.
4. **Topology is convex** if $\{\mathbf{U} : \mathbf{U}$ open and monotone $\}$ is a topological subbase.

If the space is compact ordered then the topology is convex and the space is normally ordered (see Nachbin (1965)).

For non-compact metric spaces the concept of **normally ordered spaces** with closed order is useful.

$(E, \prec)$ is **weakly normally** ordered if for each pair of disjoint compact sets $\mathbf{F}$(decreasing), $\mathbf{G}$(increasing) there exist a real function f increasing, with values in unit interval such that $f(\mathbf{F}) = 0, f(\mathbf{G}) = 1$.

Sufficient conditions for a space to be weakly normally ordered Polish space are given by Lindqvist (1988).

A metric space E can be ordered in such a way that $\prec$ and metric d are simultaneously determined by a semimetric ρ on E, for example on R ,

$$\rho(x, y) = \sup(0, x - y))$$

so that

$$x \prec y \Leftrightarrow \rho(x, y) = 0,$$

and d is equivalent to $\max\{\rho(x, y), \rho(y, x)\}$.

If such a metric exists $(E, \prec)$ is **uniformly ordered.**

Normally ordered spaces are related to uniformly ordered spaces. If $(E, \prec)$ is uniform ordered, and in addition, for every $x \in E$ and increasing $\mathbf{A}$ there exist $y \in \mathbf{A}$ such that $x \prec y$ then $(E, \prec)$ is normally ordered.

If E is countable set with the metric $d : d(x, y) = 1$ if x is not equal to y then E is normally ordered.

It is most convenient to work with compact ordered spaces. However, it is natural to use normally ordered spaces and closed order (see Lindqvist (1988)).

Orderings on $P(E)$

Next we introduce the set of all probability measures on E, denoted by $P(E)$.
For E ordered with $\prec$, several classes of sets and functions on E are introduced and using them, orderings on $P(E)$ are defined. For example
Sets:(Massey (1987))

- $S_{\prec}(E)$ increasing sets,
- $\{\{x\}\uparrow, x \in E\} \cup E$,
- $\{E \setminus \downarrow \{x\}\} \cup E$.

where a set A is increasing if $x \in A$ and $x \prec y$ implies $y \in A$. $\{x\}\uparrow = \{y : x \prec y\}$.
Functions:(Stoyan (1983))

- $I_i(E)-$ $\prec$-bounded increasing real functions,
- $I_{ic}(E)-$ $\prec$ -bounded increasing and convex functions.

If $\prec$ makes E a lattice, for example $E = \prod_i X_i$, where X_i totally ordered, then modular functions are considered (see Chang (1990)).

The following orderings are typical.

$$P_1 \prec_S P_2 \text{ iff } P_1(A) \leq P_2(A), \ A \in S_{\prec}(E),$$

$$P_1 \prec_I P_2 \text{ iff } \int_E f dP_1 \leq \int_E f dP_2, \text{ for } f \in I(E).$$

The case $\prec_{I_i} = \prec_{st}$ is called the **strong stochastic ordering.**

General information on ordered topological spaces is contained in Nachbin (1965). This theory is well developed and in consequence most of the results on ordering of probability measures on general spaces is based on ordered spaces. It should be mentioned however that there are orderings for probability measures which are not based in a natural way on the concept of ordered general spaces, for example convex orderings.

The core of the theory of stochastic orderings consists of results for compact spaces, and further generalizations are made for uniformizable ordered spaces, locally compact spaces and Polish spaces which of course are related to compactness.

General results on strong stochastic ordering

A natural element connecting two probability measures on E is a stochastic kernel $k(x,.)$. With a probability measure and a kernel we associate its convolution which is a probability measure on an appropriate product.

$$\mu * k(B_1, B_2) = \int_{B_1} k(x, B_2) d\mu(x).$$

If $k(x,.)$ has its support in $\{y : x \leq y\}$, we call it **upward.**

The fundamental and best known result for strong orderings of probability measures on general spaces is the following theorem (see e.g. Kamae et al. (1977))

Theorem A. *Suppose E is a Polish partially ordered space. The following conditions are equivalent for $\mu_1, \mu_2 \in P(E)$:*

(i)

$$\mu_1 \prec_{st} \mu_2,$$

(ii) *there exists* $\nu \in P(E^2)$ *with marginals* μ_i *and support in* $\{x \prec y\}$,

(iii) *there exists a r.v.* Z *and* $f_1, f_2 : R \to E$ *such that* $f_1 \prec f_2$ *and* $f_i(Z)$ *has distribution* μ_i ,

(iv) *there exists an upward kernel* k *on* E^2 *such that* $\mu_1 * k(E, .) = \mu_2(\,)$,

(v) $\mu_1(B) \leq \mu_2(B)$ *for all closed* $B \in I_i(E)$.

General space results for stochastic processes we divide into two cases: discrete time case, (i.e. product spaces results), and continuous time case. Other sets of indexes shall be treated separately, as for example random measures.

A fundamental technical role for these results play stochastic kernels. For a discrete time process $(\mathbf{X}(n))$ we consider the following family of kernels:

$$p_{n+1}(\mathbf{x}^n, .) = \Pr(\mathbf{X}(n+1) \in . \mid \mathbf{X}^n = \mathbf{x}^n),\ n \geq 1,$$

where $\mathbf{x}^n = (x_1, \ldots, x_n)$, and $\mathbf{X}_n \in E^n$.
For a continuous time process $(\mathbf{X}_t)$ we introduce

$$p_{t^n}(\mathbf{x}^{n-1}, .) = \Pr(\mathbf{X}_{t_n}) \in . \mid \mathbf{X}_{t^{n-1}} = \mathbf{x}^{n-1}),\ n > 1, \tag{1}$$

where $\mathbf{X}_{t^{n-1}} = (\mathbf{X}_{t_1}, \ldots, \mathbf{X}_{t_{n-1}})$, $t_1 \prec t_2 \prec \ldots \prec t_{n-1}$. In this case we call such a family a $D_E(R_+)$ family if there exists a continuous time process $(\mathbf{X}_t)$ with paths in $D_E(R_+)$, fulfilling (1).

Note that $\prec_{st}$ on $P(E)$ is a closed partial ordering with the weak convergence topology.

Stochastic ordering for random sequences i.e. processes in discrete time is characterized by finite dimensional distributions.

Theorem B. *Suppose that* $\mathbf{X}(n), \mathbf{Y}(n) \in E_n$, *for* $n \geq 1$, *and a sequence of Polish ordered spaces* (E_n). *If* $\mathbf{X}(1) \prec_{st} \mathbf{Y}(1)$ *and*

$$p_n^{\mathbf{X}}(\mathbf{x}^{n-1}, .) \prec_{st} p_n^{\mathbf{Y}}(\mathbf{y}^{n-1}, .),$$

for $\mathbf{x}^{n-1} \prec \mathbf{y}^{n-1}, n > 1$, *then*

$$(\mathbf{X}(n)) \prec_{st} (\mathbf{Y}(n)).$$

If the state space E has a compatible vector space structure i.e. addition and scalar multiplication are continuous, and $A + x \in I(E)$, for all $x \in E$ and $A \in I(E)$ then we we are able to construct two sequences with respectively ordered increments in E.

Theorem C. *If* $\mathbf{X}(n), \mathbf{Y}(n) \in E$, *for* $n \geq 1$, *and for the state space as above with*

$$p_{n+1}^{\mathbf{X}}(\mathbf{x}^n, A + x_n) = \Pr(\mathbf{X}(n+1) - \mathbf{X}(n) \in A \mid \mathbf{X}^n = \mathbf{x}^n),$$

and $p_n^{\mathbf{Y}}$ *respectively. Suppose that*

$$\mathbf{X}(1) \prec_{st} \mathbf{Y}(1)$$

and

$$p^{\mathbf{X}}_{n+1}(\mathbf{x}^n, x_n + .) \prec_{st} p^{\mathbf{Y}}_{n+1}(\mathbf{y}^n, y_n + .)$$

for all $\mathbf{x}^n$ *and* $\mathbf{y}^n$ *such that* $x_{n+1} - x_n \prec y_{n+1} - y_n, x_1 \prec y_1$.
Then

$$(\mathbf{X}(n+1) - \mathbf{X}(n)) \prec_{st} (\mathbf{Y}(n+1) - \mathbf{Y}(n)),$$

on E^∞ *with the coordinatewise ordering.*

In the continuous time case, stochastic ordering on D_E is also characterized by finite dimensional distributions. This is expressed with stochastic kernels.

Theorem D. *If* $(\mathbf{X}_t, \mathbf{Y}_t) \in D_E(I)$, $\mathbf{X}(1) \prec_{st} \mathbf{Y}(1)$ *and*

$$p^{\mathbf{X}}_{t^n}(\mathbf{x}^{n-1}, .) \prec_{st} p^{\mathbf{Y}}_{t^n}(\mathbf{y}^{n-1}, .),$$

for all $\mathbf{x}^{n-1} \prec \mathbf{y}^{n-1}$.
then

$$(\mathbf{X}_t) \prec_{st} (\mathbf{Y}_t),$$

on D_E, *with the coordinatewise ordering.*

Now we turn our attention to general orderings of Markov processes. First, we introduce a very useful notation.

Operators on P(E)

Consider operator $\mathbf{T} : P(E) \to P(E)$. We call $\mathbf{T}$ monotone if

$$P_1 \prec_I P_2 \Rightarrow \mathbf{T}P_1 \prec_I \mathbf{T}P_2.$$

We say that two such operators are comparable if

$$\mathbf{T}_1 \prec_I \mathbf{T}_2 \text{ iff } \mathbf{T}_1 P \prec_I \mathbf{T}_2 P, \text{ for all } P \in P(E).$$

We call a Markov process $(\mathbf{X}_t)$ **monotone** if the corresponding semigroup of operators $(\mathbf{P}_t)$ acting on measures are monotone.

Theorem E. (Stoyan (1983)) *Let* $(E, \prec)$ *be partially ordered Polish space with closed order ,* $\prec_I$ *be an ordering on* $P(E)$ *generated by a class of real functions* $I(E)$. *Time homogeneous Markov process* $(\mathbf{X}_t)$ *is monotone iff the corresponding operators* $(\mathbf{P}_t)$ *acting on functions fulfill:*

$$f \in I(E) \Rightarrow \mathbf{P}_t f \in I(E), \ t > 0.$$

Theorem F. *Let* $(E, \prec)$ *be as above. If* $(\mathbf{X}_t)$ *and* $(\mathbf{Y}_t)$ *are two homogeneous Markov processes with transition kernels*

$$P_{\mathbf{X}}(t, x, B) \text{ and } P_{\mathbf{Y}}(t, x, B)$$

respectively. Suppose that

$$\mathbf{X}_0 \prec_I \mathbf{Y}_0,$$

and for all $t > 0$, *and*

$$P_{\mathbf{X}}(t,x,.) \prec_I P(t,x,.) \prec_I P_{\mathbf{Y}}(t,x,.), \tag{2}$$

for some monotone $P(t,x,.)$, $t > 0$, *then*

$$\mathbf{X}_t \prec_I \mathbf{Y}_t,$$

for all $t > 0$.

Particle systems state space

Let $E = \mathbf{X} = \{0,1\}^S$, where S is a countable set. We take $\prec$ on $\mathbf{X}$ as a coordinatewise ordering, and for $\prec_I$ we take $\prec_{st}$.

The generator for a Markov process with values in $\mathbf{X}$ is usually defined by

$$\mathbf{A}f(\xi) = \sum_{x\in S} c(x,\xi)[f(\xi_x) - f(\xi)], \xi \in \mathbf{X},$$

where c denotes the intensity of the change from ξ into ξ_x , which differs only in the coordinate x.

Theorem G. (Liggett (1985)) *Suppose* S *is finite. If* $\mu_1, \mu_2 \in P(\mathbf{X})$ *with positive value on each point of* $\mathbf{X}$ *and*

$$\mu_1(\xi \wedge \vartheta)\mu_2(\xi \vee \vartheta) \geq \mu_1(\xi)\mu_2(\vartheta),$$

for all $\xi, \vartheta \in \mathbf{X}$
then

$$\mu_1 <_{st} \mu_2.$$

Product type state spaces

Let $E = \mathcal{X}_1 \times \ldots \times \mathcal{X}_n$, where $\mathcal{X}_i$ are totally ordered measure spaces with σ-finite measures σ_i. Suppose that $\mu \in P(E)$ has density f with respect to the product measure $\sigma_1 \times \ldots \times \sigma_n$.

As $\prec$ we take the coordinatewise order $\leq$. The lattice operations $\wedge$ and $\vee$ we define also coordinatewise.

We say that $f \in \mathrm{MTP}_2$ if

$$f(\mathbf{x} \wedge \mathbf{y})f(\mathbf{x} \vee \mathbf{y}) \geq f(\mathbf{x})f(\mathbf{x}).$$

If f is strictly positive then the MTP_2 property is equivalent to TP_2 in each pair of arguments.

For two densities f, g we define an order by

$$f <_{TP_2} g$$

if

$$f(\mathbf{x} \wedge \mathbf{y})g(\mathbf{x} \vee \mathbf{y}) \geq f(\mathbf{x})g(\mathbf{y}),$$

for all $\mathbf{x}, \mathbf{y} \in E$.

Theorem H. *Let E be as above. If for two time homogeneous Markov processes $(\mathbf{X}_t)$ and $(\mathbf{Y}_t), t \geq 0$ the distributions μ, ν of the initial values $\mathbf{X}_0$, $\mathbf{Y}_0$ have densities f, g such that $f <_{TP_2} g$ and*

$$P_{\mathbf{X}}(t, (\mathbf{x}, \mathbf{y})) <_{TP_2} P_{\mathbf{Y}}(t, (\mathbf{x}, \mathbf{y})),$$

on E^2, for all $(\mathbf{x}, \mathbf{y})$, $t \geq 0$, then

$$\mu \mathbf{P}_t^{\mathbf{X}} <_{TP_2} \nu \mathbf{P}_t^{\mathbf{Y}}, \ t > 0.$$

Discrete state space

Let $E = \{1, 2, \ldots\}$. Consider a time homogeneous Markov process with the intensity matrix $\mathbf{Q}$. For $\prec$ we take the natural order $\leq$.

Theorem I. *Let $\mathbf{Q}^{\mathbf{X}}$ and $\mathbf{Q}^{\mathbf{Y}}$ be two intensity matrices for Markov processes $(\mathbf{X}_t)$ and $(\mathbf{Y}_t)$. If $\mathbf{X}_0 \leq_{st} \mathbf{Y}_0$ and*

$$\sum_{u \geq v} q_{ru}^{\mathbf{X}} \leq \sum_{u \geq v} q_{su}^{\mathbf{Y}},$$

for all $r \leq s, v \leq r, v > s$,
then

$$(\mathbf{X}_t) \leq_{st} (\mathbf{Y}_t).$$

Theorem J. *Under the above assumptions $\mathbf{P}_t^{\mathbf{X}}$ is monotone $(\leq_{st})$ iff*

$$\sum_{u \geq v} q_{ru}^{\mathbf{X}} \leq \sum_{u \geq v} q_{su}^{\mathbf{X}},$$

for all $r \leq s, v \leq r, v > s$.

2.9 Stochastic ordering of point processes

The simplest description of a point process on R_+ is perhaps by a sequence of random variables $0 = T_0 < T_1 < \cdots$ on a probability space $(\Omega, \mathcal{F}, P)$. We assume that $T_n \to \infty$, as $n \to \infty$, i.e., that the process is non-explosive. Each realization of $\{T_n\}$ can be treated then as a measure on $\mathcal{R}_+$,

$$N(\omega) = \sum_{n \geq 0} \delta_{T_n(\omega)},$$

where δ_t denotes the atomic measure concentrated at t, i.e., $\delta_t(B) = 1$ if $t \in B$, and $\delta_t(B) = 0$ otherwise, for all bounded Borel sets B. Measures of this type are integer valued, and belong to the set of all Radon (i.e., locally bounded) measures on $\mathcal{R}_+$, which is equipped with the ν-topology to be a Polish space. In other words a point process can be viewed as a random measure, i.e., a random element of $\mathcal{N}$ (the space of integer valued Radon measures). This definition of a point process provides us with a "global" description of the point process. An ordering on $\mathcal{N}$, that will be used in the sequel, is defined by

$$\mu \prec_{\mathcal{N}} \nu \quad \text{if} \quad \mu(B) \leq \nu(B), \tag{1}$$

for all bounded Borel sets B, where $\mu, \nu \in \mathcal{N}$.

Restricting our attention to the sets $B = [0, t]$, $t \geq 0$, we shall consider

$$N_t = N([0,t]) = \sum_{n \geq 0} I_{(T_n \leq t)}, \quad t \geq 0,$$

where I denotes the indicator function, and we see that (N_t) has right-continuous trajectories. More formally, (N_t) is a random element of the space $\mathcal{D}[0, \infty)$ (functions which are right-continuous with left-hand limits), which is usually equipped with the Skorohod topology to be a Polish space. We call (N_t) a **counting process** since the process (N_t) counts consecutive jumps of the process with the passage of time t. Thus (N_t) gives us a "time dynamic" view of the point process. An ordering on $\mathcal{D}$, that will be used in the sequel, is defined by

$$f \prec_{\mathcal{D}} g \quad \text{if} \quad f(t) \leq g(t), \quad t \geq 0, \tag{2}$$

where $f, g \in \mathcal{D}[0, \infty)$.

A "local" description of a point process is given by the sequence of interpoint distances $X_n = T_n - T_{n-1}$, $n \geq 1$. The sequence $\{X_n\}$ is a random element in $\mathcal{R}_+^\infty$. For $\mathbf{x} = (x_1, x_2, \ldots)$ and $\mathbf{y} = (y_1, y_2, \ldots)$, such that $\mathbf{x}, \mathbf{y} \in \mathcal{R}_+^\infty$, we will consider the following ordering

$$\mathbf{x} \prec_\infty \mathbf{y} \quad \text{if} \quad x_i \leq y_i, \quad i \geq 1. \tag{3}$$

We now introduce different concepts of stochastic comparisons for point processes. Depending on which of the descriptions of point processes, given above, we adopt, we obtain different kinds of stochastic comparisons. Each time we are within the setting of partially ordered Polish spaces. Suppose that $N : \Omega \to \mathcal{N}$ and $N' : \Omega \to \mathcal{N}$ are two point processes treated as random measures. Define

$$N <_{\text{st-}\mathcal{N}} N' \quad \text{if} \quad E\phi(N) \leq E\phi(N'),$$

for all $\prec_{\mathcal{N}}$-nondecreasing real functions ϕ on $\mathcal{N}$ for which the expectations exist.
Suppose that (N_t), and (N_t') are counting processes. Define

$$N <_{\text{st-}\mathcal{D}} N' \quad \text{if} \quad E\psi((N_t)) \leq E\psi((N_t')),$$

for all $\prec_{\mathcal{D}}$-nondecreasing real functions ψ on $\mathcal{D}$ for which the expectations exist.
Suppose that N and N' are two point processes with the interpoint distances $\mathbf{X} = (T_1 - T_0, T_2 - T_1, \ldots)$ and $\mathbf{X}' = (T_1' - T_0', T_2' - T_1', \ldots)$, respectively. Define

$$N <_{\text{st-}\infty} N' \quad \text{if} \quad Ef(\mathbf{X}') \leq Ef(\mathbf{X}),$$

for all $\prec_\infty$-nondecreasing real functions f on $\mathcal{R}_+^\infty$ for which the expectations exist.

The above introduced stochastic orderings can also be characterized by means of some finite dimensional vectors as follows.

Theorem A.

(i) *$N <_{\text{st-}\mathcal{N}} N'$ if, and only if,*

$$(N(B_1),\ldots,N(B_n)) <_{\text{st}} (N'(B_1),\ldots,N'(B_n))$$

for all bounded Borel sets $B_1,\ldots,B_n$, $n \geq 1$, where $<_{\text{st}}$ denotes the usual stochastic order in $\mathcal{R}^n$;

(ii) *$N <_{\text{st-}\mathcal{D}} N'$ if, and only if,*

$$(N_{t_1},\ldots,N_{t_n}) <_{\text{st}} (N'_{t_1},\ldots,N'_{t_n}),$$

for all $t_1 < \cdots < t_n$, $n \geq 1$;

(iii) *$N <_{\text{st-}\infty} N'$ if, and only if,*

$$(X'_1,\ldots,X'_n) <_{\text{st}} (X_1,\ldots,X_n),$$

for all $n \geq 1$.

For a proof of (i) see Rolski and Szekli (1991), and of (ii), (iii) see Stoyan (1983).

Up till now we did not assume that N and N' are defined on the same probability space. However, the above orderings retain their full power for applications through the following a.s. comparison property, which follows from Strassen's theorem (see Theorem 2.8.A. (iii))

Theorem B.

(i) *$N <_{\text{st-}\mathcal{N}} N'$ if, and only if, there exist realizations of N and N' which are $\check{N}$ and $\check{N}'$, say, on the common probability space, such that*

$$P(\check{N}(B) \leq \check{N}'(B),\ B \in \mathcal{B}^1) = 1$$

($\mathcal{B}^1$ denotes the Borel σ-field);

(ii) *$N <_{\text{st-}\mathcal{D}} N'$ if, and only if, there exist realizations of N and N' which are $\check{N}$ and $\check{N}'$, say, on the common probability space, such that*

$$P(\check{N}_t \leq \check{N}'_t,\ t \geq 0) = 1;$$

(iii) *$N <_{\text{st-}\infty} N'$ if, and only if, there exist realizations of N and N' which are $\check{N}$ and $\check{N}'$, say, on the common probability space, such that for the corresponding interpoint distances $\{\check{X}_i\}$ and $\{\check{X}'_i\}$*

$$P(\check{X}_i \geq \check{X}'_i,\ i \geq 1) = 1.$$

The a.s. property in case (i) means that $\check{N}$ is a thinning of $\check{N}'$ a.s.; in case (ii) it means that $\check{N}$ has a.s. earlier and more numerous points than $\check{N}'$ before each time instant t; in case (iii) the corresponding interpoint distances are shorter for $\check{N}'$ than for $\check{N}$ a.s.. From this it is immediate that

$$N <_{\text{st-}\mathcal{N}} N' \Longrightarrow N <_{\text{st-}\mathcal{D}} N', \tag{4}$$

and that

$$N <_{\text{st-}\infty} N' \Longrightarrow N <_{\text{st-}\mathcal{D}} N'. \tag{5}$$

It is easy to see that, in general, $N <_{\text{st-}\infty} N' \not\Leftrightarrow N <_{\text{st-}\mathcal{N}} N'$. Also, in general, $N <_{\text{st-}\mathcal{N}} N' \not\Leftrightarrow N <_{\text{st-}\infty} N'$.

It is possible sometimes to construct explicitly a.s. comparable versions of point processes, which implies stochastic ordering for them, exemplifying Strassen's theorem. Consider first a single nonnegative random variable X. Let F_X be the distribution function of X and assume that it is absolutely continuous with density function f_X. Denote the survival function of X by $\overline{F}_X = 1 - F_X$. The corresponding cumulative hazard function and hazard rate function are defined as $R_X = -\log \overline{F}_X$ and $r_X = f_X/\overline{F}_X$, respectively. We have (Lemma 2.2.C.) that $R_X(x) = \int_0^x r_X(u)du$, $x \geq 0$.
Denote the left-continuous pseudo inverses of F_X and of R_X by $F_X^{-1}(u) = \inf\{x : F_X(x) \geq u\}$, $u \in (0,1)$, and $R_X^{-1}(p) = \inf\{x : R_X(x) \geq p\}$, $p \geq 0$, respectively. Then X has the representation (Lemma 1.1.D.)

$$X =^d F_X^{-1}(U),$$

where U is a uniform $(0,1)$ random variable, and $=^d$ denotes equality in distribution. From this it follows that X also has the representation (Theorem 1.2.F.)

$$X =^d R_X^{-1}(E), \tag{6}$$

where E is a standard (i.e., mean one) exponential random variable.

The idea of the representation in (6) can be used for the purpose of representing the interpoint distances $X_1, X_2, \ldots$ corresponding to a point process N. Here, let $T_0 = 0$, and denote, inductively, the jump points of N by $T_i = T_{i-1} + X_i$, $i = 1, 2, \ldots$. Let $E_1, E_2, \ldots$ be a sequence of independent standard exponential random variables. Then we have the representation

$$T_1 = X_1 =_{\text{st}} R_{X_1}^{-1}(E_1) \; (= \hat{T}_1, \text{ say}), \tag{7}$$

$$T_2 = T_1 + X_2 =_{\text{st}} \hat{T}_1 + R_{X_2|\hat{T}_1}^{-1}(E_2) \; (= \hat{T}_2, \text{ say}), \tag{8}$$

where $R_{X_2|t_1}$ is the cumulative hazard function corresponding to the conditional distribution of X_2 given that $T_1 = t_1$. And, in general, for $n \geq 1$,

$$T_n =_{\text{st}} \hat{T}_{n-1} + R_{X_n|\hat{T}_1,\ldots,\hat{T}_{n-1}}^{-1}(E_n) \; (= \hat{T}_n, \text{ say}), \tag{9}$$

where $R_{X_n|t_1,\ldots,t_{n-1}}$ is the cumulative hazard function corresponding to the conditional distribution of X_n given that $T_1 = t_1, \ldots, T_{n-1} = t_{n-1}$. From the construction described above which is the hazard construction from Chapter 2, Section 2, we have (see Theorem 2.2.L)

$$\mathbf{T} = (T_1, T_2, I) =_{\text{st}} (\hat{T}_1, \hat{T}_2, I) = \hat{\mathbf{T}}.$$

For our purposes it is convenient to represent $\hat{\mathbf{T}}$ as a function of the cumulative sums of the E_n's. Denote $P_0 = 0$ and $P_n = \sum_{i=1}^{n} E_i$, $n = 1, 2 \ldots$. Then, for $n \geq 1$, (9) can be rewritten as

$$\hat{T}_n =_{\text{st}} \hat{T}_{n-1} + R^{-1}_{X_n|\hat{T}_1,\ldots,\hat{T}_{n-1}}(P_n - P_{n-1}). \tag{10}$$

Fix an n and $0 = t_0 < t_1 < \cdots < t_n$. For $t \geq t_n$ define (regarding empty sums as zeroes)

$$a_n(t; t_0, \ldots, t_n) = R_{X_1}(t_1) + \sum_{i=2}^{n} R_{X_i|t_1,\ldots,t_{i-1}}(t_i - t_{i-1}) + R_{X_{n+1}|t_1,\ldots,t_n}(t - t_n) \tag{11}$$

($a_0(t; t_0) = R_{X_1}(t)$). Then (10) can be 'inverted' as

$$P_{n+1} = a_n(\hat{T}_{n+1}; t_0, \hat{T}_1, \ldots, \hat{T}_n), \quad n > 0. \tag{12}$$

The family of functions $\{a_n(\cdot)\}$ is called the compensator function family associated with the point process N (see Corollary 2.5.P). Roughly speaking, the function a_n describes the total hazard accumulated by the process N at time t, provided t is some time point between the nth and the $(n+1)$st jumps of the process. In the theory of point processes (12) expresses the well known fact that the transformation of a point process by its compensator results in a standard Poisson process, which is known as the random time change theorem for point processes (for a simple proof in many dimensions see, e.g., Brown and Nair (1988)). Note that for every possible realization $0 < p_1 < p_2 < \ldots$ of $P_1, P_2, \ldots$, (10) assigns a realization $0 = t_0 < t_1 < t_2 < \cdots$ of $\hat{T}_1, \hat{T}_2, \ldots$. In an abbreviated form we will then write $\mathbf{t} = \psi_{\{a_n\}}(\mathbf{p})$, where $\mathbf{t} = (t_1, t_2, \ldots)$ and $\mathbf{p} = (p_1, p_2, \ldots)$. For a more detailed description see Kwieciński and Szekli (1991).

If the functions a_n's are differentiable a.s. in the first variable then from (11) it is seen that $\frac{d}{dt}a_n(t; t_0, \ldots, t_n)$ is just the value, at time $t - t_n$, of the hazard rate of the conditional distribution of X_{n+1} given that $T_1 = t_1, \ldots, T_n = t_n$. Accordingly the (random) internal stochastic intensity of N is given by (Theorem 2.5.K)

$$\lambda(t) = r_{n+1}(t - T_n; T_1, \ldots, T_n) \quad \text{for} \quad t \in [T_n, T_{n+1}),\ n \geq 0,$$

where $r_{n+1}(t; t_1, \ldots, t_n)$ is the hazard rate function corresponding to the conditional distribution of X_{n+1} given that $T_1 = t_1, \ldots, T_n = t_n$.

It turns out that some useful sufficient criteria for stochastic ordering of point processes can be given in terms of the compensator function families of the given processes. The following two theorems (the first one is taken from Kwieciński and Szekli (1991)) are clear from the above discussion.

Theorem C. *Consider two point processes N and N' with corresponding compensator function families $\{a_n(\cdot)\}$ and $\{a'_n(\cdot)\}$. Assume that the functions in both of these families are continuous in the first variable. If*

$$a_n(t; t_0, \ldots, t_n) \leq a'_n(t; s_0, \ldots, s_n), \tag{13}$$

for all $\mathbf{t} = \psi_{\{a_n\}}(\mathbf{p})$, $\mathbf{s} = \psi_{\{a'_n\}}(\mathbf{p})$, $0 < p_1 < p_2 < \cdots$, $t \geq t_n$, then

$$N <_{\text{st-}\mathcal{D}} N'. \tag{14}$$

In particular, (14) holds whenever (13) holds for all $s_1 < \cdots < s_n$ and $t_1 < \ldots < t_n$ such that $s_i \leq t_i$, $i = 1, \ldots, n$.

It is worth mentioning, however, that it is possible that (13) holds for all $\mathbf{s}$ and $\mathbf{t}$ as in Theorem C., but not for all $\mathbf{s}$ and $\mathbf{t}$ such that $s_i \leq t_i$, $i = 1, \ldots, n$ (see, e.g., Kwieciński and Szekli (1991)).

Theorem D. *Consider two point processes N and N' with corresponding compensator function families $\{a_n(\cdot)\}$ and $\{a'_n(\cdot)\}$. Assume that the functions in both of these families are continuous in the first variable. If*

$$a_n(t + t_n; t_0, \ldots, t_n) \leq a'_n(t + s_n; s_0, \ldots, s_n),$$

for all $\mathbf{t} = \psi_{\{a_n\}}(\mathbf{p})$, $\mathbf{s} = \psi_{\{a'_n\}}(\mathbf{p})$, $p_1 < p_2 < \cdots$, $t \geq 0$, then

$$N <_{\text{st-}\infty} N'.$$

Sufficient conditions for the order $<_{\text{st-}\mathcal{N}}$ are given in the next result which is taken from Rolski and Szekli (1991).

Theorem E. *Consider two point processes N and N' with corresponding compensator function families $\{a_n(\cdot)\}$ and $\{a'_n(\cdot)\}$. Assume that the corresponding predictable stochastic intensities λ, λ' do exist. If*

$$r_{n+1}(t; t_0, \ldots, t_n) \leq r'_{n+k+1}(t; s_0, \ldots, s_{n+k}), \tag{15}$$

for all $n, k \geq 0$, $\{t_0, \ldots, t_n\} \subseteq \{s_0, \ldots, s_{n+k}\}$ and $t > s_{n+k}$ then

$$N <_{\text{st-}\mathcal{N}} N'.$$

We give here only a heuristic description of a construction of a probability space $(\Omega, \mathcal{F}, P)$ and two point processes $\check{N}$ and $\check{N}'$ on it such that all points of $\check{N}$ are points of $\check{N}'$ ($\check{N}$ is a thinning of $\check{N}'$). For a formal construction we refer to Rolski and Szekli (1991).

In the first step we select at random points according to the distribution of N' (for example on the canonical space), we denote them by $S_1, S_2, \ldots$. For each realization, which is a sequence of points on R_+, we decide, point by point, whether to accept or reject a given point, a decision depending on the history of previous decisions. We accept the point S_1 with probability r_0/r'_0, and if it is accepted, the second one with probability $r_1(S_2; t_0, S_1)/r'_2(S_2; S_0, S_1)$, otherwise $r_0/r'_2(S_2; S_0, S_1)$. Continuing this procedure, we obtain a point process, which is thinner than the first one. It is possible to compute its stochastic intensity, and to show that it is consistent with the compensator family of N, therefore from the predictability assumptions we obtain that the distribution of the thinner process is equal to the distribution of N.

Another construction is possible by so called Poisson embedding, which has nice geometric interpretations (see Lindvall (1988)).

The thinning result of Theorem E. was generalized by Last (1993), for a general treatment see Last and Brandt (1994).

PROBLEMS AND REMARKS

A. Consider two renewal processes N and N' with renewal distributions F and F', respectively, having failure rates r, r'. If $r(s) \leq r'(t)$, $t \leq s$ then $N <_{st-\mathcal{N}} N'$. Using the above thinning construction to the delayed $N_s(t) = N(s+t) - N(s)$ and the corresponding simple renewal process N with a DFR renewal distribution, N_s, N can be realized in such a way that N_s is obtained from N by deleting only a finite number of points starting from the origin. (See Section 2.10 for details) [Brown (1980), Lindvall (1986),(1988), Rolski and Szekli (1991)].

B. If N, N' are Poisson point processes with intensities λ, λ', respectively, such that $\lambda \leq \lambda'$ then $\check{N}$ can be obtained as a thinning of $\check{N}'$ (intensities can be time dependent). [Lewis and Shedler (1978)].

C. If N is a $\lambda-$ Poisson process and N' is a renewal process with a renewal distribution F having failure rate r, such that $\lambda \leq r$ then N can be realized as a thinning of N' (similarly a renewal process can be obtained as a thinning of a Poisson process with a higher intensity). [Miller (1979)].

2.10 Renewal processes

Let $\{X_n, n \geq 1\}$ ba an i.i.d. sequence of nonnegative random variables with a common distribution function F, such that $F(0) < 1$, $m_F = \int_0^\infty x dF(x) < \infty$. Letting $T_0 = 0$, $T_n = \sum_{i=1}^n X_i$, we define the **renewal counting process** by $N(t) = \sup\{n : T_n \leq t\}$. (For a process $(N(t),\ t \geq 0)$ we use abbreviate notation N). The function $M(t) = EN(t)$ is the **renewal function.** Note that this function is well defined, i.e. $M(t) < \infty$, $t \geq 0$. It is readily seen that $N(t) \geq n$ if and only if $S_n \leq t$, therefore $EN(t) = \sum_{i=1}^\infty \mathbf{Pr}(N(t) \geq n) = \sum_{i=1}^\infty \mathbf{Pr}(T_n \leq t) = \sum_{i=1}^\infty F^{n*}(t)$.

The following two theorems are classical in the renewal theory.

Theorem A. *If $N(t)$ is a renewal process then*

(i) $\lim_{t\to\infty} \frac{N(t)}{t} = \frac{1}{m_F}$ *a.s.*;

(ii) *(elementary renewal theorem)* $\lim_{t\to\infty} \frac{M(t)}{t} = \frac{1}{m_F}$.

Theorem B. *If $N(t)$ is a renewal process with a nonlattice renewal distribution F then the following statements are true and are equivalent*

(i) *(Blackwell's theorem)*

$$\lim_{t\to\infty} M(t+h) - M(t) = \frac{h}{m_F};$$

(ii) *(key renewal theorem)*

$$\lim_{t\to\infty} \int_0^t h(t-x) dM(x) = \frac{\int_0^\infty h(x)dx}{m_F},$$

for all directly Riemann integrable functions h.

We omit the case of lattice distributions, for this case and proofs see Feller (1971), Ross (1983).

If the distance from the origin to the first renewal point X_1 in a renewal process has a different distribution function G, say, then $N^d(t) = sup\{n : T_n \leq t\}$ is a **delayed renewal process.** Theorem A., and B. remain true for delayed renewal processes if we take in their formulations $M^d(t) = EN^d(t)$ in place of $M(t)$. An important special case is when X_1 is distributed accordingly to the stationary renewal distribution

$$\tilde{F}(x) = \frac{1}{m_F}\int_0^x 1 - F(u)du,$$

in this case we denote the delayed renewal process by $\tilde{N}(t)$. It is then called equilibrium (stationary) renewal process because the increments of this process form a stationary stochastic process, and $E\tilde{N}(t) = t/m_F$. Define $Z(t) = \sum_{i=1}^{N(t)+1} X_i - t$, the **forward recurrence time** at t, and $A(t) = t - \sum_{i=1}^{N(t)} X_i$, the **renewal age** at t. Denote the distribution function of $Z(s)$ by $F_{(Z(s))}(x)$. Another important case of a delayed renewal process is when X_1 has the distribution function $F_{(Z(s))}$. We denote the corresponding delayed renewal process by N_s, it is related to the nondelayed renewal process by $N_s(t) = N(t+s) - N(s)$.

It has been of interest to understand the extend to which monotonicity and aging properties of a renewal distribution are inherited in some fashion by its renewal process, and can result in obtaining bounds for the renewal function. The following theorem summarizes some results from Barlow et al.(1963) and Barlow and Proschan (1964).

Theorem C. *If N is a renewal process with a IFR renewal distribution function F then*

(i)

$$M(t) \geq kM(t/k), \; k = 1, 2, \ldots;$$

(ii)

$$M(t+h) - M(t) \geq M(h), \; t, h \geq 0;$$

(iii)

$$M(t) \leq \frac{t}{m_F}\frac{F(t)}{\tilde{F}(t)} \leq \frac{t}{m_F};$$

(iv)

$$tf(0) \leq M(t) \leq R(t),$$

where f is a density of F and $R(t)$ is the total hazard rate function of F.

If F is DFR then the inequalities in the above theorem $(i) - (iii)$ are reversed.

For NBU distributions we have the following (see Marshall and Proschan (1972))

Theorem D. *If N is a renewal process with a renewal distribution function F then*

(i) *F is NBU iff $\mathcal{L}_{N(t)} * \mathcal{L}_{N(s)} <_{st} \mathcal{L}_{N(t+s)}$, $t, s \geq 0$, ($\mathcal{L}_{N(t)}$ denotes the law of $N(t)$);*

(ii) *If F is NBU then $VarN(t) \leq EN(t)$;*

(iii) *If F is NBU then $N(t) <_{st} N_s(t)$, $s,t \geq 0$.*

The conditions $(i),(iii)$ can be modified for NWU distributions. From (i) it follows that for NBU distributions $M(s+t) \geq M(s) + M(t)$. The condition (iii) is equivalent to the property that $F_{(Z(s))} <_{st} F$, $s \geq 0$, which is implied by $F \in NBU$, equivalent to $F_s <_{st} F$. (F_s is the residual life distribution function defined in Section 1.4). The class defined by the condition $\tilde{F} <_{st} F$ is called $NBUE$ (with the reversed relation $NWUE$). For $NBUE$ class we have

Theorem E. *If N is a renewal process with a renewal distribution F then*

(i) *F is NBUE iff $N(t) <_{st} \tilde{N}(t)$, $t \geq 0$;*

(ii) *If F is NBUE then $M(t) \leq \frac{t}{m_F}$, $t \geq 0$.*

Again, the above theorem can be modified for $NWUE$ distributions.

Much stronger monotonicity results can be obtained for DFR distributions. In order to formulate results in the strongest version we turn back to the description of renewal processes by their stochastic intensities (see Example 1 after Corollary 2.5.P.). For a non-delayed renewal process we have the corresponding compensator function family given by

$$a_n(t; t_0, \ldots, t_n) = R_{X_1}(t - t_n) + R_{X_1}(x_n) + \cdots + R_{X_1}(x_1), \tag{1}$$

for $n \geq 1$ and $t \geq t_n$, where $x_n = t_n - t_{n-1}$. And if R_{X_1} is differentiable then

$$\lambda(t) = r_{X_1}(t - T_n) \quad \text{for} \quad t \in [T_n, T_{n+1}),\ n \geq 0.$$

The next result identifies pairs of non-delayed renewal processes which are ordered in accordance with $<_{\text{st-}\mathcal{D}}$ and $<_{\text{st-}\infty}$.

Theorem F. *Consider two non-delayed renewal processes N and N' with renewal distribution functions F and F', respectively. The following conditions are equivalent.*

(i) *$F' <_{\text{st}} F$,*

(ii) *$N <_{\text{st-}\mathcal{D}} N'$,*

(iii) *$N <_{\text{st-}\infty} N'$.*

Indeed, from the independence of the interpoint distances it follows that (i)$\Longleftrightarrow$(iii). From (5) it follows that (iii)$\Longrightarrow$(ii). The implication (ii)$\Longrightarrow$(i) is obvious.

In order to obtain $N <_{\text{st-}\mathcal{N}} N'$, which, for non-delayed renewal processes, is a stronger result than $N <_{\text{st-}\infty} N'$, one can use the following theorem. It follows at once from Theorem E..

Theorem G. *Consider two non-delayed renewal processes N and N' with renewal distribution functions F and F', respectively. Let r and r' denote the hazard rate functions corresponding to F and F', respectively. If $r(t) \leq r'(s)$ for all $0 \leq s \leq t$ then*

$$N <_{\text{st-}\mathcal{N}} N'.$$

Remark. For any two distribution functions F and F' denote $F' <_{\rm h} F$ if $r(t) \le r'(t)$ for all $t \ge 0$ (see Section 1.4.). The condition $r(t) \le r'(s)$ for all $0 \le s \le t$ is fulfilled if in addition to $F' <_{\rm h} F$ we have that F is DFR (i.e., $r(t)$ is nondecreasing in t) or that F' is DFR.

Comparison of a non-delayed renewal process with a delayed one (which has the same interrenewal distribution), or a comparison of two delayed renewal processes with different delays (but with the same interrenewal distribution) will be useful.

Theorem H. *Consider two delayed renewal processes N^d and $N^{d'}$, with the corresponding delay distributions F^d and $F^{d'}$ and with the same interrenewal distribution after the delay. The following statements are equivalent.*

(i) $F^{d'} <_{\rm st} F^d$,

(ii) $N^d <_{\text{st-}\mathcal{D}} N^{d'}$,

(iii) $N^d <_{\text{st-}\infty} N^{d'}$.

Theorem I. *Consider two delayed renewal processes N^d and $N^{d'}$, with the corresponding delay distributions F^d (with hazard rate function r^d) and $F^{d'}$ and with the same interrenewal distribution F (with hazard rate function r) after the delay. If $F^{d'} <_{\rm h} F^d$ and $r^d(t) \le r(s)$ for all $0 \le s \le t$ then*

$$N^d <_{\text{st-}\mathcal{N}} N^{d'}.$$

Remark. The condition $r^d(t) \le r(s)$ for all $0 \le s \le t$ is fulfilled, for example, if $F <_{\rm h} F^d$ and F is DFR, or if $F <_{\rm h} F^d$ and F^d is DFR.

As a special case of Theorem I. we obtain:

Corollary J. *Consider a non-delayed renewal process N and a delayed one N^d (with delay distribution F^d with hazard rate function r^d) with the same interrenewal distribution F (with hazard rate function r). If $r^d(t) \le r(s)$ for all $0 \le s \le t$ then*

$$N^d <_{\text{st-}\mathcal{N}} N.$$

Remark. Again, the condition $r^d(t) \le r(s)$ for all $0 \le s \le t$ is fulfilled, for example, if $F <_{\rm h} F^d$ and F is DFR, or if $F <_{\rm h} F^d$ and F^d is DFR.

As an application of Corollary J. we will derive now a result of Brown (1980) and some other strong monotonicity properties.

Theorem K. *If N is a renewal process with a DFR renewal distribution F then*

(i) $N_u <_{\text{st-}\mathcal{N}} N_s,\ u \ge s \ge 0$;

(ii) $M(t)$ *is concave;*

(iii) $\lim_{t\to\infty} M(t+h) - M(t) = \frac{h}{m_F}$ *in a decreasing way.*

Proof. If at the time zero a component has its initial age $a \geq 0$ then the survival function of its life is given by the residual life distribution $\overline{F}_a(t) = \overline{F}(a+t)/\overline{F}(a)$, $t \geq 0$. Consider the case in which the initial age a is random and equals to $A(s)$, for a fixed $s \geq 0$. That is, let the delay distribution (the distribution of X_1) be $F_{(Z(s))}$, and consider the corresponding delayed renewal process N_s. If we assume that F is DFR then $F <_{\mathrm{h}} F_a$ for all $a \geq 0$. Thus, $F <_{\mathrm{h}} F_{(Z(s))}$ (see e.g. Rolski and Szekli (1991)). Therefore the assumptions in the remark after Corollary J. are fulfilled for $F^d = F_{(Z(s))}$, and thus we obtain

$$N_s <_{\text{st-}\mathcal{N}} N \quad \text{for all} \quad s \geq 0, \tag{2}$$

From Theorem 1.9.A.(i)) this is equivalent to

$$(N(B_1+s),\ldots,N(B_n+s)) <_{\mathrm{st}} (N(B_1),\ldots,N(B_n)) \quad \text{for all} \quad s \geq 0, \tag{3}$$

or equivalent to

$$(N(B_1+u),\ldots,N(B_n+u)) <_{\mathrm{st}} (N(B_1+s),\ldots,N(B_n+s)) \quad \text{for all} \quad s \leq u, \tag{4}$$

for the corresponding numbers of points in the shifted Borel sets $B_1,\ldots,B_n$ (see Lindvall (1988) and Rolski and Szekli (1991)), thus

$$N_u <_{\text{st-}\mathcal{N}} N_s \quad \text{for all} \quad u \geq s \geq 0,$$

which completes the proof of (i), (ii), (iii) can be obtained by taking expectations in (4) for $n = 1$. □

The conditions (ii), and (iii) of Theorem K. were first proved by Brown (1980), and next by Lindvall (1986), Hansen (1991), Rolski and Szekli (1991) using different approaches. It is Brown's conjecture that $M(t)$ concave implies F is DFR. This conjecture was proved false for discrete type distributions by Shanthikumar (1986), but remains open in general. It is worth mentioning here that in the DFR case $\lambda(t) = r(A(t))$ is stochastically nonincreasing in t (in fact, $A(t)$, $Z(t)$ are nondecreasing in t in the hazard rate order; see Kijima (1992)).

The following bounds were proposed by Brown (1980)

Theorem L. *If F is DFR such that $\psi(a) = \int_0^\infty exp(at)dF(t) < \infty$ for some $a > 0$ then*

$$U(t) - \frac{c(F,a)}{exp(at)-1} \leq M(t) \leq U(t), \; t > 0,$$

where $U(t) = \frac{t}{m_F} + \frac{m_F[2]}{2m_F^2} - 1$, $c(F,a) = \frac{1}{am_F} - \frac{m_F[2]}{2m_F^2} - \frac{1}{\psi(a)-1}$, $m_F[2]$ denotes the second moment of F.

The above result holds for a broader class of renewal distributions $IMRL$ (defined for example by $\tilde{F} \in DFR$), see Brown (1980) for improved bounds when F is DFR.

PROBLEMS AND REMARKS

A. If N^d is a delayed renewal process then $\frac{N^d(t)}{t} \to \frac{1}{m_F}$ *a.s.*, $t \to \infty$.

B. $\lim_{t\to\infty} \mathbf{Pr}(A(t) \le x) = \lim_{t\to\infty} \mathbf{Pr}(Z(t) \le x) = \tilde{F}(x)$.

C. If F is NBU then $F_{(Z(s))} <_{st} F$, $s \ge 0$.

D. If $F_{(Z(s))} <_{st} F$ for all $s \ge 0$ then F is $NBUE$.

E. $F_{(Z(s))} <_{st} F_{(Z(t))}$, $0 \le s \le (\ge) t$ iff $M(t)$ is concave (convex). [Shaked and Zhu (1991)].

F.

(1) If $M(t)$ is concave then F is NWU;

(2) If $M(t)$ is convex then F is IFR;

(3) If $M(t)$ has a logconcave density then F has logconcave density.

[Hansen (1990a,1990b)].

2.11 Comparison of replacement policies

Age replacement policy is perhaps the simplest type of preventive maintenance policy where a renewal replacement policy is modified. Here items are replaced at failure (unplanned replacements) or replaced when the age of a component reaches some fixed value T (planned replacements). Denote by $0 = T_0^A < T_1^A < \cdots$ the consecutive failure times under an age replacement policy, with a fixed replacement age T. The failure point process for this age replacement policy is

$$N_t^{A,T} = \sum_{n\ge 0} I_{\{T_n^A \le t\}}.$$

Note that $(N_t^{A,T})$ is a renewal process for each T. If r denotes the failure rate of the interpoint distances in the corresponding renewal process (remember that $N^{A,T}$ is a modification of a renewal process N) then the failure rate function of the interpoint distances for $(N_t^{A,T})$ is given by

$$r^T(t) = \sum_{k=0}^{\infty} r(t - kT) I_{[kT,(k+1)T)}(t), \; t \ge 0. \tag{1}$$

The corresponding cumulative hazard function is given by

$$R^T(t) = \sum_{k=0}^{\infty} (R(t - kT) + kR(T)) I_{[kT,(k+1)T)}(t), \; t \ge 0. \tag{2}$$

Thus the compensator function family of $(N_t^{A,T})$ is given by

$$a_0^{A,T}(t; 0) = R^T(t), \; t \ge 0,$$

and

$$a_n^{A,T}(t; t_0, \ldots, t_n) = R^T(t - t_n) + R^T(x_n) + \cdots + R^T(x_1), \tag{3}$$

for $n \geq 1$ and $t \geq t_n$, where $x_n = t_n - t_{n-1}$ and R^T is given by (2). The corresponding internal stochastic intensity is

$$\lambda^{A,T}(t) = r(t - T_n - kT) \quad \text{for} \quad t \in [T_n + kT, (T_n + (k+1)T) \wedge T_{n+1}),$$
$$k \in \{m \geq 0 : T_n + mT \leq T_{n+1}\}, \ n \geq 0.$$

Block replacement policy is another type of modification of the renewal replacement policy. Items are replaced at failure (unplanned replacements), and also at fixed times $T, 2T, \ldots$ (planned replacements). Denote by $0 = T_0^B < T_1^B < \cdots$ the consecutive failure times under a block replacement policy, with a fixed replacement block T. The failure point process for this block replacement policy is

$$N_t^{B,T} = \sum_{n \geq 0} I_{\{T_n^B \leq t\}}.$$

In general $(N_t^{B,T})$ need not be a renewal process. The compensator function family of $(N_t^{B,T})$ is given by

$$a_0^{B,T}(t) = R^T(t), \ t \geq 0,$$

and

$$a_n^{B,T}(t; t_0, \ldots, t_n) = R_{t_n}^{B,T}(t - t_n) + a_{n-1}^{B,T}(t_n; t_0, \ldots, t_n), \ t \geq t_n, \tag{4}$$

where

$$R_{t_n}^{B,T}(t) = R(t) I_{[0,\delta_n)}(t) + (R^T(t - \delta_n) + R(\delta_n)) I_{[\delta_n,\infty)}(t), \ t \geq 0,$$

for $\delta_n = \min\{(kT - t_n) > 0 : k = 1, 2, \ldots\}$ (R^T is given in (2)). The corresponding internal stochastic intensity is, for $n \geq 0$,

$$\lambda^{B,T}(t) = \begin{cases} r(t - T_n), & \text{if } t \in [T_n, K_n T); \\ r(t - kT), & \text{if } t \in [kT, (k+1)T \wedge T_{n+1}), \ k \in \{m \geq K_n : mT \leq T_{n+1}\}, \end{cases}$$

where $K_n = \min\{k \geq 1 : kT > T_n\}$.

Minimal repair replacement policy is a policy in which an item, upon failure, is only minimally repaired, i.e., the item is brought back to a working condition, but it is only as good then as it was just before it failed. Formally, if an item, with life distribution F, fails at some time t then the survival function corresponding to the distribution of the next failure time of the item is given by $\overline{F}(s)/\overline{F}(t)$, $s \geq t$. Denote by $0 = T_0^M < T_1^M < \cdots$ the consecutive failure times under a minimal repair replacement policy. The failure point process for this minimal repair replacement policy is

$$N_t^M = \sum_{n \geq 0} I_{\{T_n^M \leq t\}}.$$

Obviously (N_t^M) is a nonhomogeneous Poisson process with mean function $R = -\log \overline{F}$. Thus, the compensator function family of (N_t^M) is given by

$$a_n^M(t; t_0, I, t_n) = R(t), \tag{5}$$

for $n \geq 1$ and $t \geq t_n > \cdots > t_0 = 0$, and the intensity function of (N_t^M) is given by

$$\lambda^M(t) = \frac{d}{dt} R(t), \quad t > 0.$$

Apart from the basic age, block and minimal repair replacement policies, there are many other replacement policies that have been discussed in the literature. For example, Sumita and Shanthikumar (1988) and Block, Langberg and Savits (1993) have introduced and studied several classes of replacement policies. A recent review paper is by Beichelt (1993).

We turn our attention to comparison results among various replacement policies involving the $<_{\text{st-}\infty}$ ordering. We already know that this ordering is stronger than the $<_{\text{st-}\mathcal{D}}$ ordering and it allows us to compare infinite sequences of interfailure moments. Most of the results formulated in the literature in this area are concerned with the $<_{\text{st-}\mathcal{D}}$ ordering which allows comparison of the counting processes of failures over arbitrary time intervals $[0, t]$. An advantage of having some comparison results in terms of $<_{\text{st-}\infty}$ relies on the fact that they enable us in addition to compare (increasing) functionals defined on the whole sequence of interfailure times, for the corresponding replacement policies. Some benefit and cost functions have this structure.

Theorem A. *Consider a non-delayed renewal process N) and the family of failure point processes for age replacement policies $N^{A,T}$, $T \geq 0$, associated with it. Then*

(i) *F is NBU if, and only if, $N^{A,T} <_{\text{st-}\infty} N$ for all $T \geq 0$,*

(ii) *F is NWU if, and only if, $N^{A,T} >_{\text{st-}\infty} N$ for all $T \geq 0$.*

Proof. (i) Let F be NBU. This is equivalent to the fact that $R(t)$ is superadditive, i.e.,

$$R(x + y) \geq R(x) + R(y), \ x, y \geq 0.$$

To prove that $N^{A,T} <_{\text{st-}\infty} N$, for all $T \geq 0$, it is enough (see Theorem 2.10.F.) to show that $F <_{\text{st}} F^T$, where F^T denotes the distribution function of interfailure times in $(N_t^{A,T})$. This in turn is equivalent to

$$R^T(t) \leq R(t), \ t \geq 0, \tag{6}$$

for the corresponding cumulative hazard functions. Recalling the expression for R^T (see (2)) we have, for $t \in [kT, (k+1)T)$, $k \geq 0$, that

$$\begin{aligned} R^T(t) &= R(t - kT) + kR(T) \\ &\leq R(t) - R(kT) + kR(T) \\ &\leq R(t), \end{aligned}$$

where the two inequalities follow from the superadditivity of R. Thus (6) holds. The converse implication in (i) follows from the above inequalities and the fact that T is arbitrary. The proof of (ii) is similar. □

The next result can be proven in a similar manner. We omit the detailed proof.

Theorem B. *Consider a non-delayed renewal process N and the family of failure point processes for age replacement policies $N^{A,T}$), $T \geq 0$, associated with it. Then*

(i) *F is NBU if, and only if, $N^{A,T} <_{\text{st-}\infty} N^{A,mT}$ for all $T \geq 0$ and positive integers m,*

(ii) *F is NWU if, and only if, $N^{A,T} >_{\text{st-}\infty} N^{A,mT}$ for all $T \geq 0$ and positive integers m.*

For block replacement policies we have the following result.

Theorem C. *Consider a non-delayed renewal process N and the family of failure point processes for block replacement policies $N^{B,T}$), $T \geq 0$, associated with it. Then*

(i) *F is NBU if, and only if, $N^{B,T} <_{\text{st-}\infty} N$ for all $T \geq 0$,*

(ii) *F is NWU if, and only if, $N^{B,T} >_{\text{st-}\infty} N$ for all $T \geq 0$.*

Proof. We apply Theorem 2.9. D. to prove (i). Assume that F is NBU, i.e., that R is superadditive. It is enough to show that

$$a_n(t+t_n; t_0, \ldots, t_n) \geq a_n^{B,T}(t+s_n; s_0, \ldots, s_n),$$

for all $\mathbf{t} = \psi_{\{a_n\}}(\mathbf{p})$ and $\mathbf{s} = \psi_{\{a_n^{B,T}\}}(\mathbf{p})$, $p_1 < p_2 < \cdots$, $t \geq 0$, $n \geq 0$. Take an arbitrary $\mathbf{p}$, and let $\mathbf{t} = \psi_{\{a_n\}}(\mathbf{p})$ and $\mathbf{s} = \psi_{\{a_n^{B,T}\}}(\mathbf{p})$. From (1) and (4) we have

$$a_n(t+t_n; t_0, \ldots, t_n) = R(t) + a_{n-1}(t_n; t_0, \ldots, t_{n-1})$$

and

$$a_n^{B,T}(t+s_n; s_0, \ldots, s_n) = R_{s_n}^{B,T}(t) + a_{n-1}^{B,T}(s_n; s_0, \ldots, s_{n-1}).$$

Note that from the definition of $\mathbf{t}$ and $\mathbf{s}$ we have

$$a_{n-1}(t_n; t_0, \ldots, t_{n-1}) = a_{n-1}^{B,T}(s_n; s_0, \ldots, s_{n-1}) = p_n.$$

Hence it is enough to show that

$$R(t) \geq R_{s_n}^{B,T}(t), \; n \geq 0, \; t \geq 0.$$

For $n = 0$ this means

$$R(t) \geq R^T(t), \; t \geq 0,$$

which is true by (6). For $n \geq 1$ it remains to verify that (see (4))

$$R(t) \geq R(t) I_{[0,\delta_n)}(t) + (R^T(t-\delta_n) + R(\delta_n)) I_{[\delta_n,\infty)}(t), \; t \geq 0,$$

where here $\delta_n = \min\{(kT - s_n) > 0 : k = 1, 2, \ldots\}$. For $t \in [0, \delta_n)$ we have equality. For $t \geq \delta_n$ we should have

$$R(t) \geq R^T(t-\delta_n) + R(\delta_n),$$

or, equivalently

$$R(x+\delta_n) - R(\delta_n) \geq R^T(x),$$

for $x \geq 0$. Now, from the supermodularity of R we have that $R(x+\delta_n) \geq R(x) + R(\delta_n)$. Therefore it is enough to show that

$$R(x) \geq R^T(x), \; x \geq 0,$$

which again holds by (6). The converse implication in (i) follows from the fact that the times to the first failure, in $N^{B,T}$ and in N, are then stochastically ordered, which gives the formula (6) for arbitrary $T \geq 0$, and this implies that F is NBU. The proof of (ii) is similar. □

Let $\mathcal{Z} = \{z_n\}$ be a sequence of real numbers satisfying $0 < z_1 < z_2 < \cdots$ with $z_n \to 0$. By a block replacement policy with block schedule $\mathcal{Z}$ we mean a policy in which planned replacements by a new item occur at the pre assigned times $z_1, z_2, \ldots$. We denote the point process corresponding to the unplanned replacements in this policy by $(N_t^{B,\mathcal{Z}})$. Note that the classical block replacement policy corresponds to the choice $z_n = nT$, $n \geq 1$.

Theorem D. *Consider a non-delayed renewal process N and the family of failure point processes for block replacement policies $N^{B,\mathcal{Z}}$ with block schedules $\mathcal{Z}$, associated with it. Then*

(i) *F is NBU if, and only if, $N^{B,\mathcal{Z}} <_{\text{st-}\infty} N$ for all block schedules $\mathcal{Z}$,*

(ii) *F is NWU if, and only if, $N^{B,\mathcal{Z}} >_{\text{st-}\infty} N$ for all block schedules $\mathcal{Z}$.*

Theorem 2.9.D. can also be used in order to prove the following result.

Theorem E. *Consider the processes $N^{A,T}$, and $N^{B,T}$, which are associated with the same non-delayed renewal process N, and having the same fixed value $T \geq 0$. If F is IFR (DFR) then*

$$N^{B,T} <_{\text{st-}\infty} N^{A,T} \quad (N^{A,T} <_{\text{st-}\infty} N^{B,T}).$$

Proof. From Theorem 2.10.D., it is enough to show, for F which is IFR, that

$$a_n^{B,T}(t + t_n; t_0, \ldots, t_n) \leq a_n^{A,T}(t + s_n; s_0, \ldots, s_n),$$

for $\mathbf{t} = \psi_{\{a_n^{B,T}\}}(\mathbf{p})$, $\mathbf{s} = \psi_{\{a_n^{A,T}\}}(\mathbf{p})$, and arbitrary $\mathbf{p}$, with $p_1 < p_2 < \cdots$, $t \geq 0$, $n \geq 0$. For $\mathbf{t}$ and $\mathbf{s}$ as above we have

$$a_{n-1}^{B,T}(t_n; t_0, \ldots, t_{n-1}) = a_{n-1}^{A,T}(s_n; s_0, \ldots, s_{n-1}).$$

Thus (2) and (4) imply that it is enough to show that

$$R^T(t) \geq R_{t_n}^{B,T}(t), \; t \geq 0, \; n \geq 1$$

(for $n = 0$ we have equality). Equivalently, substituting the expression for $R_{t_n}^{B,T}$ (see (4)), it is seen that it is enough to show that

$$R^T(t) \geq R(t) I_{[0,\delta_n)}(t) + (R^T(t - \delta_n) + R(\delta_n)) I_{[\delta_n,\infty)}(t).$$

This trivially holds for $t \in [0, \delta_n)$ because δ_n <T. Thus we aim at verifying that, for $n \geq 1$ (we substitute $x = t - \delta_n$),

$$R^T(x + \delta_n) \geq R^T(x) + R(\delta_n), \; x \geq 0,$$

or, equivalently, that

$$R^T(x+\delta_n) - R^T(x) \geq R(\delta_n), \ x \geq 0.$$

Since R is convex the increments of R^T on intervals of the length δ_n ($\delta_n < T$) are greater (or equal if $x = kT$, $k \geq 0$) than $R(\delta_n)$ (increments for convex functions are nondecreasing). This completes the proof for the IFR case. The proof for the DFR case is similar. □

Another result that we can easily deduce from Theorem 2.10.F. is the following.

Theorem F. *Consider the processes $N^{A,T}$, and $N^{A,T'}$, which are associated with the same non-delayed renewal process N, but having different replacement ages T and T'. Then*

(i) *If F is IFR then $N^{A,T} <_{\text{st-}\infty} N^{A,T'}$ whenever $T \leq T'$,*

(ii) *If F is DFR then $N^{A,T'} <_{\text{st-}\infty} N^{A,T}$ whenever $T \leq T'$.*

Proof. (i) Because both $N^{A,T}$ and $N^{A,T'}$ are renewal processes it is enough to show that $R^T(t) \leq R^{T'}(t)$, $t \geq 0$. This inequality is evident from the fact that R is convex, and from the definition of R^T (see (2)). The proof of (ii) is similar. □

We now turn our attention to the minimal repair replacement policy (N_t^M). From Theorem 2.9.D. and (5) we obtain the following result (here we denote by $N^{M,F}$ the point process corresponding to a minimal repair replacement policy with the underlying life distribution F).

Theorem G. *Consider the processes $N^{M,F}$ and $N^{M,F'}$. If $F' <_{\text{disp}} F$ then $N^{M,F} <_{\text{st-}\infty} N^{M,F'}$.*

The next comparison follows from 2.10. (1) and (5).

Theorem H. *Let N be a non-delayed renewal process and let N^M be the point process corresponding to the minimal repair replacement policy which is associated with N.*

(i) *If the underlying life distribution F is NBU then $N <_{\text{st-}\infty} N^M$.*

(ii) *If the underlying life distribution F is NWU then $N >_{\text{st-}\infty} N^M$.*

Proof. (i) The compensator function family of N is described in 2.10. (1) and the compensator function family of N^M is described in (5). Thus, by Theorem 2.9.D. it is enough to show that, for $n \geq 1$ and $t > t_n > \cdots > t_1 > 0$,

$$R(t-t_n) + R(t_n - t_{n-1}) + \cdots + R(t_1) \leq R(t),$$

where $R = -\log \overline{F}$. But this follows from the superadditivity of R (which is the NBU assumption). The proof of (ii) is similar. □

As a corollary of Theorems D. and H. we obtain the following result.

Corollary I. *Consider the processes $N^{B,Z}$ and N^M which have the same underlying life distribution F. If F is NBU (NWU) then $N^{B,Z} <_{\text{st-}\infty} N^M$ ($N^{B,Z} >_{\text{st-}\infty} N^M$) for all block schedules Z.*

We now prove some comparison results in terms of the $<_{\text{st-}\mathcal{N}}$ ordering. Comparisons of two failure processes corresponding to different replacement policies with this ordering result in the following interpretation: At any arbitrary (small) time interval the smaller process has stochastically less failures than the larger process. (One policy is "totally" worse than the other.) This situation is intimately connected with the DFR property of the corresponding lifetime.

Theorem J. *Consider a non-delayed renewal process N and the failure process of an age policy $N^{A,T}$ associated with it, for a fixed replacement age T. If F is DFR then*

$$N <_{\text{st-}\mathcal{N}} N^{A,T}.$$

Proof. The processes N and $N^{A,T}$ are both renewal processes. By assumption, F is DFR, hence from Theorem 2.10. G. and the remark after it, it is enough to show that $F^T <_{\text{h}} F$, which is evident since in this case the assumption that F is DFR implies that $r(t) \leq r^T(t)$, $t \geq 0$ (see (1)). □

Theorem K. *Consider a non-delayed renewal process N_t and the failure processes $N^{B,T}$ and $N^{B,mT}$ corresponding to two block replacement policies which are associated with it, for a fixed T and a fixed positive integer m.*

(i) *If F is IFR then $N^{B,mT} >_{\text{st-}\mathcal{N}} N^{B,T}$.*

(ii) *If F is DFR then $N^{B,mT} <_{\text{st-}\mathcal{N}} N^{B,T}$.*

Proof. (ii) From Theorem 2.9. E. and (4) it is enough to verify that

$$\frac{d}{dt}R^{B,mT}(t-t_n) \leq \frac{d}{dt}R^{B,T}(t-s_{n+k}), \tag{7}$$

for all $n,k \geq 0$, $\{t_0,\ldots,t_n\} \subseteq \{s_0,\ldots,s_{n+k}\}$ and $t > s_{n+k}$. Note that for r^T and r^{mT} (using (1) and the assumption that F is DFR we have $r^{mT}(t) \leq r^T(t)$, $t \geq 0$, because $\{imT,\ i \geq 0\} \subseteq \{iT,\ i \geq 0\}$. Denote by δ the distance from s_{n+k} to the time of the next planned replacement in $\{imT,\ i \geq 0\}$, which is ImT, say. It is clear that, for $t > s_{n+k} + \delta = ImT$, we have $\frac{d}{dt}R^{B,T}(t-t_n) = r^T(t-ImT)$ and $\frac{d}{dt}R^{B,mT}(t-s_{n+k}) = r^{mT}(t-ImT)$. Thus we have the required inequality (7) for $t > ImT$.

Now let $t \in (s_{n+k}, ImT]$. Denote by ε the distance from s_{n+k} to the time of the last planned replacement in $\{imT,\ i \geq 0\}$ before s_{n+k}, which is JmT, say. Then

$$\begin{aligned}
\frac{d}{dt}R^{B,mT}(t-t_n) &= r(t-t_n \vee JmT) \\
&\leq r(t-s_{n+k}) \\
&\leq r^T(t-s_{n+k}) \\
&= \frac{d}{dt}R^{B,T}(t-s_{n+k}),
\end{aligned}$$

where the first inequality follows from the DFR assumption and the fact that $s_{n+k} \geq t_n \vee JmT$, and the second inequality follows from the DFR assumption and (1). This establishes (7) also for $t \in (s_{n+k}, ImT]$. The proof of (i) is similar. □

The idea of the proof of Theorem K. can actually be used for the purpose of proving the next, stronger result. First we need the following definition. Let $\mathcal{Z}$ and $\mathcal{Z}'$ be two block schedules. We say that $\mathcal{Z}'$ is a **refinement** of $\mathcal{Z}$ (in the language of Shaked and Shanthikumar (1989), $\mathcal{Z}$ is said to be **thinner** than $\mathcal{Z}'$, or $\mathcal{Z}'$ is said to be **denser** than $\mathcal{Z}$) if $\mathcal{Z} \subseteq \mathcal{Z}'$.

Theorem L. *Consider a non-delayed renewal process N and the failure processes $N^{B,\mathcal{Z}}$ and $N^{B,\mathcal{Z}'}$, corresponding to two block replacement policies with block schedules $\mathcal{Z}$ and $\mathcal{Z}'$, which are associated with it. Suppose that $\mathcal{Z} \subseteq \mathcal{Z}'$.*

(i) *If F is IFR then $N^{B,\mathcal{Z}} >_{\text{st-}\mathcal{N}} N^{B,\mathcal{Z}'}$.*

(ii) *If F is DFR then $N^{B,\mathcal{Z}} <_{\text{st-}\mathcal{N}} N^{B,\mathcal{Z}'}$.*

For minimal repair replacement policies we have the following result as a consequence of Theorem 2.9.E. and (5).

Theorem M. *Consider the processes $N^{M,F}$ and $N^{M,F'}$. If $F' <_{\text{h}} F$ then $N^{M,F} <_{\text{st-}\mathcal{N}} N^{M,F'}$*

The following policy, which is a generalization of a minimal repair replacement policy and a block replacement policy with block schedule $\mathcal{Z}$, is studied in Block, Langberg and Savits (1990a). In this policy, upon a failure, the item is minimally repaired, and at fixed scheduled times, $\mathcal{Z} = \{z_1, z_2, \ldots\}$, planned replacements take place. Denote by $(N_t^{M,F,\mathcal{Z}})$ the point process associated with the unplanned replacements in this policy, where F is the underlying lifetime distribution. Note that between each pair of consecutive scheduled replacement times z_{n-1} and z_n the process increments itself as a nonhomogeneous Poisson process with mean function $R = -\log \overline{F}$. Thus the following result is immediate from Theorem M..

Theorem N. *Consider the processes $N^{M,F,\mathcal{Z}}$ and $N^{M,F',\mathcal{Z}}$. If $F' <_{\text{h}} F$ then $N^{M,F,\mathcal{Z}} <_{\text{st-}\mathcal{N}} N^{M,F',\mathcal{Z}}$ for all block schedules $\mathcal{Z}$.*

We will now compare the point process $N^{B,\mathcal{Z}'}$ (corresponding to a block replacement policy with block schedule $\mathcal{Z}'$) with the point process $N^{M,\mathcal{Z}}$ (corresponding to a minimal repair replacement policy with planned replacement schedule $\mathcal{Z}$), where we drop the designation F of the common life distribution which underlies both processes.

Theorem O. *Consider the processes $N^{B,\mathcal{Z}'}$ and $N^{M,\mathcal{Z}}$ with the underlying life distribution F. Suppose that $\mathcal{Z} \subseteq \mathcal{Z}'$.*

(i) *If F is IFR then $N^{M,\mathcal{Z}} >_{\text{st-}\mathcal{N}} N^{B,\mathcal{Z}'}$.*

(ii) *If F is DFR then $N^{M,\mathcal{Z}} <_{\text{st-}\mathcal{N}} N^{B,\mathcal{Z}'}$.*

Proof. (i) Let z_{n-1} and z_n be two consecutive scheduled replacement times in the thinner schedule $\mathcal{Z}$. On the interval $[z_{n-1}, z_n)$ the process $N^{M,\mathcal{Z}}$ increments itself as a nonhomogeneous Poisson process with intensity function $r = \frac{d}{dt}R$, where $R = -\log \overline{F}$. Fix a $t \in [z_{n-1}, z_n)$. Let $z'_k = \max\{z' \in \mathcal{Z}' : z' \leq t\}$ be the last scheduled replacement time in $\mathcal{Z}'$ before t. From the assumption $\mathcal{Z} \subseteq \mathcal{Z}'$ it follows that $z_{n-1} \leq z'_k \leq t < z_n$. The failure intensity of $N^{M,\mathcal{Z}}$ at time t is $r(t - z_{n-1})$, whereas (using the assumption that F is IFR the failure intensity of $N^{B,\mathcal{Z}'}$ at time t is at most $r(t - z'_k)$. Again, from the assumption that F is IFR it follows that $r(t - z'_k) \leq r(t - z_{n-1})$. Therefore from Theorem 2.9. E. we obtain $N^{B,\mathcal{Z}'} <_{st\text{-}\mathcal{N}} N^{M,\mathcal{Z}}$. The proof of (ii) is similar. □

For more results on comparison of replacement policies see e.g. Shaked and Szekli (1994) or the book of Shaked and Shanthikumar (1994).

2.12 Stochastically monotone networks

We assume that $\mathbf{X} = (\mathbf{X}(t), t \geq 0)$ is a Markov process on $E = \mathbf{N}^n$, with transition rates

$$q(\mathbf{x},\mathbf{y}) = \lim_{t\to 0} t^{-1} \Pr(\mathbf{X}(t) = \mathbf{y} \mid \mathbf{X}(0) = \mathbf{x}), \qquad \mathbf{x} \neq \mathbf{y}.$$

For convenience we define $q(\mathbf{x},\mathbf{x}) = 0$. We adopt the standard assumption

$$q(\mathbf{x}) = \sum_{\mathbf{y}\in E} q(\mathbf{x},\mathbf{y}) < \infty, \ \mathbf{x} \in E.$$

The process $\mathbf{X}$ is assumed to describe the movement or migration of discrete units in a n-node stochastic network, such as flexible manufacturing, telecommunication, computer, maintenance, distribution or biological network, or a system of n populations (for details see e.g. Serfozo (1990a)). By its nature a network is a system of interacting nodes in which the operation of a node may depend on what is happening throughout the network. A fairly wide class of Markovian network processes with system dependent transition rates which contains the Jackson processes and most of its generalizations developed up to date was introduced by Serfozo (1990a).

A natural multivariate process for modeling jointly the sizes of populations such as the queue lengths in a stochastic network is the compound birth-death process, or more general the migration process defined in Serfozo (1990a). His results on steady-state distributions utilized explicitly the reversibility of the processes. Because we do not need this property in our context, we consider systems with a simpler structure which however resembles births, deaths, and migrations of customers. Therefore we shall use the notion of generalized (multidimensional) birth-death processes and generalized migration processes.

We state now the definitions of these processes. A generic state of $\mathbf{X}$ is the vector $\mathbf{x} = (x_1, \ldots, x_n)$, where x_j denotes the number of units at node j. Units move around the nodes, possibly in bulks. Whenever the process $\mathbf{X}$ is in state $\mathbf{x}$, a typical transition will transfer it to a state $\mathbf{x} + \mathbf{a}\text{-}\mathbf{d}$. Here $\mathbf{a} = (a_1, \ldots, a_n)$ and $\mathbf{d} = (d_1, \ldots, d_n)$ denote the numbers of arrivals to and departures from the respective nodes $(\mathbf{a}, \mathbf{d} \geq \mathbf{0})$. In some cases, $\mathbf{a}$ or $\mathbf{d}$ might be zero vector. We denote the transition rates of the process $\mathbf{X}$ by

$$q(\mathbf{x},\mathbf{y}) = \begin{cases} \Lambda(\mathbf{x},\mathbf{a},\mathbf{d}) & if \ \mathbf{x},\mathbf{y} = \mathbf{x} + \mathbf{a} - \mathbf{d} \in E, \ \mathbf{a} \neq \mathbf{d} \\ 0 & otherwise. \end{cases} \tag{1}$$

Here Λ is a nonnegative function. One can represent a variety of networks by appropriate choices of $(\mathbf{a}, \mathbf{d})$, for instance networks with single unit movements, open and closed networks with splitting or merging of units (see Serfozo (1990)). An important special case of the process $\mathbf{X}$ is given by the only nonzero transition rates

$$q(\mathbf{x},\mathbf{y}) = \begin{cases} \beta(\mathbf{x},\mathbf{a}) & if\ \mathbf{x},\mathbf{y} = \mathbf{x}+\mathbf{a} \in E\ for\ some\ \mathbf{a} \\ \delta(\mathbf{x},\mathbf{d}) & if\ \mathbf{x},\mathbf{y} = \mathbf{x}-\mathbf{d} \in E\ for\ some\ \mathbf{d}. \end{cases} \tag{2}$$

Here $\beta's$ are the birth rates, $\delta's$ are the death rates, when the population is in state $\mathbf{x}$. Births do not occur at the same time as deaths.

A Markov process with transition rates (2) is called **generalized birth-death** process. A Markov process with transition rates (1) is a **generalized migration** process.

A generalized migration process with one at a time movements of size one, which are births, deaths or migration between two nodes is called a **simple birth-death-migration process** ($b-d-m$ process).

By a **Jackson network** we mean an open network of exponential single server queues under FCFS discipline, with state-dependent service rates. The queues are numbered $1,\ldots,n$. Customers arrive in independent Poisson processes with rates $\lambda r(0,j)$ at node $j, j = 1,\ldots,n$, where $\lambda > 0, r(0,j) \geq 0, \sum_{j=1}^{n} r(0,j) = 1$. Their routing is Markovian, governed by a matrix $R = (r(i,j), 1 \leq i \leq n, 0 \leq j \leq n)$. Setting $r(0,0) = 0$ we assume that $(r(i,j), 0 \leq i,j \leq n)$ constitutes an irreducible Markov chain. If at node j there are x_j customers present, including the one in service, the service rate is $\mu_j(x_j)$; we set $\mu_j(0) = 0$. We assume the usual independence assumptions (see Kelly (1979)). The Jackson network process $\mathbf{X}$ with state space $E = \mathbf{N}^n$ is a generalized migration process with the following transition rates: (we denote by $\mathbf{e}_j$ the j-th unit vector with 1 in entry j and 0 elsewhere)

$$\begin{aligned} q(\mathbf{x},\mathbf{x}+\mathbf{e}_j) &= \lambda r(0,j), \\ q(\mathbf{x},\mathbf{x}-\mathbf{e}_j+\mathbf{e}_i) &= r(j,i)\mu_j(x_j), \qquad x_j \geq 1, \\ q(\mathbf{x},\mathbf{x}-\mathbf{e}_j) &= r(j,0)\mu_j(x_j), \qquad x_j \geq 1, \end{aligned}$$

and $q(\mathbf{x},\mathbf{x}') = 0$ for all other states.
It represents the joint queue length evolution over time.

By a **Gordon-Newell** network we mean a closed network with $N \geq 1$ customers cycling. The node structure is the same as in a Jackson network, the routing of the customers is Markovian, governed by an irreducible stochastic matrix $R = (r(i,j), 1 \leq i,j \leq n)$. The Gordon-Newell network process $\mathbf{X}$ with state space $E = \{(x_1,\ldots,x_n) : x_j \in \mathbf{N}, j = 1,\ldots,n, x_1+\ldots+x_n = N\}$ is a generalized migration process, describing the joint queue length vector with the following transition rates:

$$q(\mathbf{x},\mathbf{x}-\mathbf{e}_j+\mathbf{e}_i) = r(j,i)\mu_j(x_j), \qquad x_j \geq 1,$$

and $q(\mathbf{x},\mathbf{x}') = 0$ for all other states.

Denote the set of feasible movements of the process $\mathbf{X}$ by F, i.e.

$$F = \{(\mathbf{a},\mathbf{d}) : \Lambda(\mathbf{x},\mathbf{a},\mathbf{d}) > 0 \text{ for some } \mathbf{x} \in E\}.$$

We assume that F is finite. Define

$$F_{\mathbf{x}} = \{(\mathbf{a}, \mathbf{d}) : \mathbf{x} + \mathbf{a} - \mathbf{d} \in E\}, \qquad \mathbf{x} \in E.$$

By $\prec_{st}$ we denote the strong stochastic order generated by a generic partial order $\prec$.

Theorem A. *A generalized birth-death process* $\mathbf{X}$ *is* $\prec_{st}$*-monotone if its possible transitions are such that:*
for each $\mathbf{x} \prec \mathbf{y}$*, and all* $(\mathbf{a}, \mathbf{d}) \in F$
i)

if $(\mathbf{a}, \mathbf{0}) \in F_{\mathbf{x}} \cap F_{\mathbf{y}}$ *and* $\neg(\mathbf{x} + \mathbf{a} \prec \mathbf{y})$ *then* $\beta(\mathbf{x}, \mathbf{a}) \leq \beta(\mathbf{y}, \mathbf{a})$
if $(\mathbf{0}, \mathbf{d}) \in F_{\mathbf{x}} \cap F_{\mathbf{y}}$ *and* $\neg(\mathbf{x} \prec \mathbf{y} - \mathbf{d})$ *then* $\delta(\mathbf{x}, \mathbf{d}) \geq \delta(\mathbf{y}, \mathbf{d})$.

ii)

if $(0, \mathbf{d}) \notin F_{\mathbf{x}}$ *and* $\neg(\mathbf{x} \prec \mathbf{y} - \mathbf{d})$ *then* $\delta(\mathbf{y}, \mathbf{d}) = 0$
if $(\mathbf{a}, 0) \notin F_{\mathbf{y}}$ *and* $\neg(\mathbf{x} + \mathbf{a} \prec \mathbf{y})$ *then* $\beta(\mathbf{x}, \mathbf{a}) = 0$,

where $\neg$ *denotes the negation.*

For results on stochastic monotonicity for simple birth-death processes see Ridder (1987), where a special case of the above proposition is proved.

For simple b-d-m processes on a state space with the coordinatewise ordering the next result yields a special useful case of Massey's (1989) general characterization result.

Theorem B. *A generalized migration process* $\mathbf{X}$ *is* $\prec_{st}$*-monotone if its possible transitions are such that:*
for each $\mathbf{x} \prec \mathbf{y}$*, and all* $(\mathbf{a}, \mathbf{d}) \in F$
i)

if $(\mathbf{a}, \mathbf{d}) \in F_{\mathbf{x}} \cap F_{\mathbf{y}}$ *and* $\neg(\mathbf{x} + \mathbf{a} - \mathbf{d} \prec \mathbf{y})$ *then* $\Lambda(\mathbf{x}, \mathbf{a}, \mathbf{d}) \leq \Lambda(\mathbf{y}, \mathbf{a}, \mathbf{d})$,
if $(\mathbf{a}, \mathbf{d}) \in F_{\mathbf{x}} \cap F_{\mathbf{y}}$ *and* $\neg(\mathbf{x} \prec \mathbf{y} + \mathbf{a} - \mathbf{d})$ *then* $\Lambda(\mathbf{x}, \mathbf{a}, \mathbf{d}) \geq \Lambda(\mathbf{y}, \mathbf{a}, \mathbf{d})$,

ii)

if $(\mathbf{a}, \mathbf{d}) \notin F_{\mathbf{x}}$ *and* $\neg(\mathbf{x} \prec \mathbf{y} + \mathbf{a} - \mathbf{d})$ *then* $\Lambda(\mathbf{y}, \mathbf{a}, \mathbf{d}) = 0$
if $(\mathbf{a}, \mathbf{d}) \notin F_{\mathbf{y}}$ *and* $\neg(\mathbf{x} + \mathbf{a} - \mathbf{d} \prec \mathbf{y})$ *then* $\Lambda(\mathbf{x}, \mathbf{a}, \mathbf{d}) = 0$.

Proof. Will be given later in this section.

For the coordinatewise ordering $\leq$ and a simple $b - d - m$ process with homogeneous transition rates the assumptions i)-ii) are fulfilled, hence the first statement of the following corollary is obvious. The second follows by direct checking the conditions.

Corollary C. *Suppose that* $\mathbf{X}$ *is a Jackson or Gordon-Newell network. If the service rates are queue length independent or non-decreasing in the queue length, then* $\mathbf{X}$ *is* $\leq_{st}$ *monotone.*

Whitt (1981) investigated open feed forward systems with respect to the partial sum order.

Recall the definition of the partial sum order. For two vectors $\mathbf{x}, \mathbf{y} \in E$ of the form $\mathbf{x} = (x_1, \ldots, x_n)$, $\mathbf{y} = (y_1, \ldots, y_n)$, we define **partial sum** order by

$$\mathbf{x} \leq_* \mathbf{y} \text{ if } \sum_{i=1}^{j} x_i \leq \sum_{i=1}^{j} y_j, \qquad j = 1, \ldots, n.$$

Under the above assumptions $\leq_*$ is a closed partial order. Whitt's result can be generalized

Theorem D. *A generalized migration process* $\mathbf{X}$ *is* $\leq_{st*}$ *-monotone if its possible transitions are such that:*
for each $\mathbf{x} \leq_* \mathbf{y}$*, and all* $(\mathbf{a}, \mathbf{d}) \in F$
i)

if $(\mathbf{a}, \mathbf{d}) \in F_{\mathbf{x}} \cap F_{\mathbf{y}}$ *and* $\neg(\mathbf{x} + \mathbf{a} - \mathbf{d} \leq_* \mathbf{y})$ *then* $\Lambda(\mathbf{x}, \mathbf{a}, \mathbf{d}) \leq \Lambda(\mathbf{y}, \mathbf{a}, \mathbf{d})$,
if $(\mathbf{a}, \mathbf{d}) \in F_{\mathbf{x}} \cap F_{\mathbf{y}}$ *and* $\neg(\mathbf{x} \leq_* \mathbf{y} + \mathbf{a} - \mathbf{d})$ *then* $\Lambda(\mathbf{x}, \mathbf{a}, \mathbf{d}) \geq \Lambda(\mathbf{y}, \mathbf{a}, \mathbf{d})$,

ii)

if $(\mathbf{a}, \mathbf{d}) \notin F_{\mathbf{x}}$ *and* $\neg(\mathbf{x} \leq_* \mathbf{y} + \mathbf{a} - \mathbf{d})$ *then* $\Lambda(\mathbf{y}, \mathbf{a}, \mathbf{d}) = 0$
if $(\mathbf{a}, \mathbf{d}) \notin F_{\mathbf{y}}$ *and* $\neg(\mathbf{x} + \mathbf{a} - \mathbf{d} \leq_* \mathbf{y})$ *then* $\Lambda(\mathbf{x}, \mathbf{a}, \mathbf{d}) = 0$.

An interesting special case now is

Corollary E. (Daduna and Szekli (1992)) *Suppose that* $\mathbf{X}$ *is a Jackson network with a routing matrix* $\mathbf{R} = \{r(i,j), 1 \leq i,j \leq n\}$ *such that* $r(i,j) > 0$ *implies that* $j = i + 1$ *or* $j = i - 1$*, and* $r(i, 0) > 0$ *if and only if i=n. If the state dependent service rates are non -decreasing at each node in the queue length then* $\mathbf{X}$ *is* $\leq_{st*}$ *-monotone.*

In view of the above corollary it is natural to ask: is it possible to find conditions on the topological structure of a network which guarantee monotonicity of the describing process independently of how the arrival and service rates are chosen? To be more precise: the routing probabilities $r(i, j)$ generate a directed connection graph with nodes $1, \ldots, n$ and an edge from i to j (in this direction) if and only if $r(i, j) > 0$. Which are the connection graphs that guarantee all networks with this underlying structure to be monotone?

We consider here the case of $\leq_*$ - monotonicity. From the above considerations it follows that allowing service rates decreasing in the queue length would destroy monotonicity in general. Excluding this case we have the surprising reversal of the last corollary.

Theorem F. (Daduna and Szekli (1992)) *The network is* $\leq_{st*}$ *monotone only if it has the connection graph of a series system as given in Corollary E., for all possible arrival rates and service rates which are increasing in queue lengths. (The numbering of the nodes has to be exactly that as prescribed in this corollary.)*

Let μ be an arbitrary distribution of $\mathbf{X}(0)$. Consider the marked point process $\mathbf{N}_\mu$on $\mathbf{R}_+ \times F$ defined by

$$\mathbf{N}_\mu(A \times B) = \sum_{t \in A} \mathbf{I}_{(\mathbf{X}(t) - \mathbf{X}(t_-) \in B)}, \;\; B \subseteq F, \;\; A \subseteq \mathbf{R}_+.$$

This point process records the times at which the Markov process $\mathbf{X}$ jumps, and "marks" at each jump time τ_i, which are the actual movement directions. We have (Lemma 2.7.E) for $\mathbf{N}_\mu$, and each B

$$\mathbf{E}[\mathbf{N}_\mu((t, t + dt) \times B) \mid \mathbf{X}(t)] = \Big(\sum_{(\mathbf{a}, \mathbf{d}) \in B} \Lambda(\mathbf{X}(t), \mathbf{a}, \mathbf{d})\Big) dt$$

This is a consequence of Levy's formula. From this we conclude that each $\mathbf{N}_\mu((0, t) \times B)$ has a stochastic intensity with respect to the internal history of $\mathbf{N}_\mu(. \times F)$.

Denote the corresponding counting processes $\mathbf{N}_\mu((0,t]\times\{(\mathbf{a},\mathbf{d})\}$ by $N_\mu(t,(\mathbf{a},\mathbf{d}))$. Recall that F is finite. For the family of point processes $\{N_\mu(t,(\mathbf{a},\mathbf{d})),(\mathbf{a},\mathbf{d})\in F\}$ denote by

$$\mathbf{T}^{(\mathbf{a},\mathbf{d})} = (T_1^{(\mathbf{a},\mathbf{d})}, T_2^{(\mathbf{a},\mathbf{d})},\ldots), \qquad T_1^{(\mathbf{a},\mathbf{d})} < T_2^{(\mathbf{a},\mathbf{d})} < \ldots,$$

its consecutive jump times. Let

$$\mathbf{T} = (\mathbf{T}^{(\mathbf{a},\mathbf{d})}, (\mathbf{a},\mathbf{d})\in F).$$

$\mathbf{T}$ is an equivalent description of the process $\mathbf{X}$, in terms of point processes, since

$$\mathbf{X}(t) = \mathbf{X}(0) + \sum_{(\mathbf{a},\mathbf{d})\in F} (\mathbf{a}-\mathbf{d})N_\mu(t,(\mathbf{a},\mathbf{d})). \tag{3}$$

Denote by $\mathbf{J}$ the set of increasing sequences in $\mathbf{R}_+^\infty$, which represent jump points of point processes. In the proof of Theorem B. we shall use a representation of the vector $\mathbf{T}$ by a transformation of a vector of standard Poisson processes given by the following lemma.

Lemma G. *Suppose that $\Pi_0 = (\pi^1,\ldots,\pi^{|F|})$ is a vector of independent standard Poisson processes. For each initial distribution μ there exist a measurable function Φ_μ^* : $\mathbf{J}^{|F|}\to\mathbf{J}^{|F|}$, such that*

$$\Phi_\mu^*(\Pi_0) =^d \mathbf{T}$$

Proof. In order to simplify the notation we proceed with an argument for a general multidimensional simple point process

$$\mathbf{N} = (N_1,\ldots,N_k).$$

By $N_i(t)$ we denote the corresponding counting processes. Each N_i is determined through its jump times

$$\mathbf{T}^i = (T_1^i, T_2^i,\ldots).$$

Denote the vector of the coordinate point processes by

$$\mathbf{T} = (\mathbf{T}^1,\ldots,\mathbf{T}^k).$$

All jumps of $\mathbf{T}$ enumerated in increasing order, we denote by

$$0=\tau_0<\tau_1<\tau_2<\ldots, \text{ and assume that } \tau_n\to\infty.$$

Let

$$\mathcal{F}_t^i = \sigma(N_i(s), s\le t), \text{ and } \mathcal{G}_t = \mathcal{G}_0 \vee \bigvee_{i=1}^k \mathcal{F}_t^i.$$

where $\mathcal{G}_0$ is an arbitrary σ-field.
The mark at n-th jump point τ_n we denote by $Z_n, n\ge 0$. The σ-field $\mathcal{G}_0$ usually describes the behavior of the initial value Z_0.
Assume that $\mathbf{N}$ has compensator $\mathbf{A} = (A_1,\ldots,A_k)$, such that $A_i(\infty)=\infty$, and realizations of $A_i(t)$ are continuous a.s.

From Corollary 2.5.P.

$$A_i(t) = \sum_{n\geq 0} a_n^i(t; \tau_0, Z_0, \ldots, \tau_n, Z_n)\mathbf{I}_{\{\tau_n \leq t < \tau_{n+1}\}},$$

where $\mathcal{R}_a = \{a_n^i(.), i = 1, \ldots, k, n \geq 0\}$ is a family of multidimensional measurable functions.

Define a transformation $\Phi_a : \mathbf{J}^k \to \mathbf{J}^k$ by the following procedure:
for $\mathbf{t}^i \in \mathbf{J}$, where $\mathbf{J}$ is as defined before and $\mathbf{t} = (\mathbf{t}^1, \ldots, \mathbf{t}^k)$, and z_0 fixed, set

$$\Phi_{z_0}(\mathbf{t}) = \mathbf{p} = (\mathbf{p}^1, \ldots, \mathbf{p}^k),$$

where for each $i = 1, \ldots, k$,

$$p_1^i = a_0^i(t_1^i; \tau_0, z_0, \ldots, \tau_{n_1^i}, z_{n_1^i}),$$

for an index n_1^i, and a sequences $(\tau_0, \tau_2, \ldots), (z_1, z_2, \ldots)$ which are determined by the increasing enumeration of $\mathbf{t}$, so that $t_1^i = \tau_{n_1^i}$;

$$p_2^i = a_1^i(t_2^i; \tau_0, z_0, \ldots, \tau_{n_2^i}, z_{n_2^i}),$$

for $t_2^i = \tau_{n_2^i}$;

$$\vdots$$

$$p_j^i = a_{j-1}^i(t_j^i; \tau_0, z_0, \ldots, \tau_{n_j^i}, z_{n_j^i}),$$

for $t_j^i = \tau_{n_j^i}$;

$$\vdots$$

Let

$$a_n^{*i}(p; .) = \inf\{t : a_n^i(t; .) \geq p\}.$$

Define $\Phi_{z_0}^* : \mathbf{J}^k \to \mathbf{J}^k$ by the following procedure:
for an arbitrary $\mathbf{p} \in \mathbf{J}$, fixed z_0, and $\tau_0 = 0$ set

$$\Phi_{z_0}^*(\mathbf{p}) = \mathbf{t},$$

where

$$t_1^{i_1} = \min\{a_0^{*i}(p_1^i; \tau_0, z_0), i = 1, \ldots, k\},$$

for an index i_1 for which the minimum is achieved.
We take $\tau_1 = t_1^{i_1}$, and $z_1 = i_1$.
In the next step we search an index for which the following minimum is attained

$$\min\{a_1^{*i}(p_1^i; \tau_0, z_0, \tau_1, z_1), i \neq i_1\} \wedge a_1^{*i}(p_2^{i_1}; \tau_0, z_0, \tau_1, z_1),$$

which is denoted by i_2. Take the value of this minimum as $t_1^{i_2}$,if $i_2 \neq i_1$, or $t_2^{i_1}$, if $i_1 = i_2$. We take $\tau_2 = t_1^{i_2}$, if $i_2 \neq i_1$, or $\tau_2 = t_2^{i_1}$, otherwise, and $z_2 = i_2$. We continue this procedure to obtain next τ_n and z_n, which in turn determine $\mathbf{t}$.

Now if $Z_0 = z_0$ is fixed, and $\Pi_0 = (\mathbf{P}^1, \ldots, \mathbf{P}^k)$ is a vector of independent standard Poisson processes then it is enough to prove that $\Phi^*_{z_0}(\Pi_0) =^d \mathbf{T}$
Suppose that for some $\mathbf{t} = (\mathbf{t}^1, \ldots, \mathbf{t}^k)$, and $\mathbf{z} = (z_1, \ldots), (z_0$ fixed), we have

$$a^i_j(t^i_j - \epsilon; \tau_0, z_0, \ldots, \tau_{n^i_j}, z_{n^i_j}) < a^i_j(t^i_j; \tau_0, z_0, \ldots, \tau_{n^i_j}, z_{n^i_j}), \tag{4}$$

for all $j \geq 1$ and $\epsilon > 0, i = 1, \ldots, k$.

For such $\mathbf{t}$ and $\tau, \mathbf{z}$ it follows

$$\Phi^*_{z_0}(\Phi_{z_0}(\mathbf{t})) = \mathbf{t}.$$

Indeed, by induction, let $\check{\mathbf{t}} = \Phi^*_{z_0}(\Phi_{z_0}(\mathbf{t}))$, fix arbitrary $i \in \{1, \ldots, k\}$, and assume that $\check{t}^i_j = t^i_j, j = 1, \ldots, n-1, (n \geq 1)$. From (4),

$$\check{t}^i_n = a^{*i}_{n-1}(a^i_{n-1}(t^i_n; \tau_0, z_0, \ldots, \tau_{n^i_{n-1}}, z_{n^i_{n-1}}); \tau_0, z_0, \ldots, \tau_{n^i_{n-1}}, z_{n^i_{n-1}}) =$$

$$= \inf\{t : a^i_{n-1}(t; \tau_0, z_0, \ldots, \tau_{n^i_{n-1}}, z_{n^i_{n-1}}) \geq a^i_{n-1}(t^i_n; \tau_0, z_0, \ldots, \tau_{n^i_{n-1}}, z_{n^i_{n-1}})\} = t^i_n.$$

Now we will see that (4) holds $\mathbf{P}^{\mathbf{T}}$ a.s., for all $n \geq 1$, rational $\epsilon > 0$, and $i = 1, \ldots, k$. ($\mathbf{P}^{\mathbf{T}}$ denotes the distribution of $\mathbf{T}$). This can be seen as follows. Fix $i \in \{1, \ldots, k\}, \epsilon > 0$, and let

$$C^i_\epsilon(t) = \mathbf{I}_{\{a^i_{N_i(t-)}(t-\epsilon; \tau_0, Z_0, \ldots, \tau_{n^i_{N_i(t-)}}, Z_{n^i_{N_i(t-)}}) = a^i_{N_i(t-)}(t; \tau_0, Z_0, \ldots, \tau_{n^i_{N_i(t-)}}, Z_{n^i_{N_i(t-)}})\}}.$$

Since both $a^i_{N_i(t-)}(t; .)$ and $a^i_{N_i(t-)}(t - \epsilon; .)$ are left continuous and adapted to $\mathcal{G}_t$, so they are predictable and $C^i_\epsilon(t)$ is also predictable.
Now

$$\mathbf{P}(a^i_j(T^i_j; \tau_0, Z_0, \ldots, \tau_{n^i_j}, Z_{n^i_j}) = a^i_j(T^i_j - \epsilon; \tau_0, Z_0, \ldots, \tau_{n^i_j}, Z_{n^i_j})) =$$

$$\mathbf{E}C^i_\epsilon(T^i_j) = \mathbf{E}\int_0^\infty C^i_\epsilon(t) d\mathbf{I}_{(T^i_j \leq t)} \leq \mathbf{E}\int_0^\infty C^i_\epsilon(t) dN_i(t) = \mathbf{E}\int_0^\infty C^i_\epsilon(t) dA_i(t) = 0.$$

Since (4) holds $\mathbf{P}^{\mathbf{T}}$ a.s., from the first part of the proof

$$\Phi^*_{z_0}(\Phi_{z_0}(\mathbf{T})) = \mathbf{T} \ a.s.$$

The proof is completed by the time change theorem (see e.g. Brown and Nair (1988)), from which we have that $\Phi_{z_0}(\mathbf{T}) =^d \Pi_0$. □

Proof. (of Theorem B.) For $\prec_{st}$-monotonicity it is enough to show that for each $\mathbf{x}, \mathbf{y} \in E$, such that $\mathbf{x} \prec \mathbf{y}$ we can construct two migration processes $\mathbf{X_x}, \mathbf{X_y}$, say on the same probability space one of which starts from $\mathbf{x}$, the second one from $\mathbf{y}$ fulfilling for each fixed $t > 0$

$$\mathbf{Pr}(\mathbf{X}_x(t) \prec \mathbf{X}_y(t)) = 1. \tag{5}$$

Take arbitrary $\mathbf{x}_0 \prec \mathbf{y}_0$, and Π_0 from Lemma G.. Trajectory by trajectory of Π_0 we obtain

$$\Phi^*_{\mathbf{y}_0}(\Pi_0), \Phi^*_{\mathbf{x}_0}(\Pi_0)$$

where the indexes $\mathbf{x}_0, \mathbf{y}_0$ of Φ^*'s indicate deterministic initial measures ($\mu = \delta_{\mathbf{x}_0}, \mu = \delta_{\mathbf{y}_0}$). From Lemma G. these are versions of point processes representing the Markov processes from formula (3) , which start from $\mathbf{x}_0$, and $\mathbf{y}_0$, respectively. We denote these versions by $\mathbf{X}^0_{\mathbf{x}_0}, \mathbf{X}^0_{\mathbf{y}_0}$, respectively. We see from the definition of $\Phi's$ and the assumptions i), ii) that these versions need not possess the property $\mathbf{X}^0_{\mathbf{x}_0}(t) \prec \mathbf{X}^0_{\mathbf{y}_0}(t)$ for all trajectories of Π_0. However they possess this property up to the random time

$$R_1 = \inf_n \{\tau_n : \Delta_{(\mathbf{a},\mathbf{d})}(\tau_n)\Delta_{(\mathbf{a},\mathbf{d})}(\tau_{n-}) < 0, \text{ for some } (\mathbf{a},\mathbf{d}) \in F\}$$

where $\Delta_{(\mathbf{a},\mathbf{d})}(t) = \Lambda(\mathbf{X}^0_{\mathbf{x}_0}(t), \mathbf{a}, \mathbf{d}) - \Lambda(\mathbf{X}^0_{\mathbf{y}_0}(t), \mathbf{a}, \mathbf{d})$, and $(\tau_n, n \geq 1)$ are the consecutive jump times of $(\mathbf{X}^0_{\mathbf{x}_0}, \mathbf{X}^0_{\mathbf{y}_0})$. They remain ordered all the time when after the consecutive jumps (τ_n) the intensities do not change the direction of their order. At time R_1 the change takes place for the first time. At this time we "reset" our construction by taking a new vector Π_1 and the initial conditions

$$\mathbf{x}_1 = \mathbf{X}^0_{\mathbf{x}_0}(R_1-), \mathbf{y}_1 = \mathbf{X}^0_{\mathbf{y}_0}(R_1-),$$

for which of course we still have $\mathbf{x}_1 \prec \mathbf{y}_1$.
Roughly speaking the nature of the construction (which follows from the definition of Φ^*) guarantees that under the assumptions in i), which regulate the behavior of the process inside the state space, and in ii) which do the same on the boundary of the state space, the coordinates of $(\mathbf{X}^0_{\mathbf{x}_0}, \mathbf{X}^0_{\mathbf{y}_0})$ jump in directions in which they remain ordered if they are ordered at the start position. The consecutive jumps of the underlying simple Poisson process are "inverted" by picewise linear functions, which depend on the jump intensities, to obtain the consecutive jumps of the process. Only a relative change of "speed" of the inversion may spoil such a behavior and this is the reason why we "reset" the construction at such a moment.
From the strong Markov property of the process $(\mathbf{X}^0_{\mathbf{x}_0}, \mathbf{X}^0_{\mathbf{y}_0})$, by continuing the process after R_1, which is a stopping time for this process we obtain new versions , $\mathbf{X}^1_{\mathbf{x}_0}, \mathbf{X}^1_{\mathbf{y}_0}$ say, of $\mathbf{X}_{\mathbf{x}_0}, \mathbf{X}_{\mathbf{y}_0}$ generated now jointly by Π_0 and Π_1. Now we continue this procedure by defining next $R_j's$ to obtain versions which are ordered for each trajectory on the first arbitrary finite number of jumps of the process, since $R_n \geq \tau_n$ a.s. $n \geq 1$. From this we finally deduce (5). □

PROBLEMS AND REMARKS

A. Let $(\mathbf{X}(t))$ be a Jackson network process with constant service rates, and $(\mathbf{Y}(t))$ a system of the equal number of parallel nodes with the same arrival and service rates acting independently. If $\mathbf{Pr}(\mathbf{X}(0) \in A) \leq \mathbf{Pr}(\mathbf{Y}(0) \in A)$ then $\mathbf{Pr}(\mathbf{X}(t) \in A) \leq \mathbf{Pr}(\mathbf{Y}(t) \in A)$, $t > 0$, for $A = \{\mathbf{y} : \mathbf{x} \leq \mathbf{y}\}$, $\mathbf{x} \in Z^n$. [Massey (1987)].

B. Consider a queueing system consisting of K identical queues in parallel. Arrivals are routed to one of the K queues by a sequence of i.i.d. Bernoulli random variables with parameter $\mathbf{p} = (p_1, \ldots, p_K)$, buffers are infinite and there is no blocking. The expected total amount of work in the system at time t and the response time of every customer are minimized (the response time in the sense of $< icx$) when $\mathbf{p} = (1/K, \ldots, 1/K)$. [Chang (1992)].

C. Consider a system in which a single Poisson stream is to be routed to N identical service stations that operate independently with service rates $\phi(x_j)$ for each j (service rates are dependent on the number of customers x_j at node j). Assume that ϕ is nondecreasing, concave and bounded. The policy *route to the shortest queue* (RSQ) is optimal for this system in the following sense. If $\mathbf{Pr}(\mathbf{X}^{RSQ}(0) \in A) \leq \mathbf{Pr}(\mathbf{X}(0) \in A)$ then $\mathbf{Pr}(\mathbf{X}^{RSQ}(t) \in A) \leq \mathbf{Pr}(\mathbf{X}(t) \in A)$, where $\mathbf{X}(t)$ is the vector of the number of customers at service stations at time t, rearranged in the decreasing order, and A is a set with increasing and Schur convex indicator function. [Menich and Serfozo (1991)].

D. Consider a system with independent Poisson input of rate λ into each of the N service stations, and a single exponential server who, when placed at a service station, serves at rate μ. The policy *serve the longest queue* (SLQ) is optimal for this system in the sense as described in the above problem. [Menich and Serfozo (1991)].

2.13 Queues with MR arrivals

We use the symbol MR/GI/1 for a G/GI/1 , FCFS queue with a Markov renewal (MR) arrival process, which we define as follows. Consider a sequence of arrays of independent positive random variables $\mathbf{X} = \{[X_{ij}^{(k)}]_{i,j\geq 1},\ k = 1, 2, \cdots\}$. For each i, j the sequence $\{X_{ij}^{(k)},\ k = 1, 2, \cdots\}$ is i.i.d. with a common distribution function F_{ij}. Let $\mathbf{Z} = \{Z_0, Z_1, \cdots\}$ be a Markov chain with the state space $\{1, 2, \cdots\}$, transition probabilities

$$a_{ij} = P(Z_k = j \mid Z_{k-1} = i),\ k \geq 1, \tag{1}$$

and an initial distribution $a_i = P(Z_0 = i)$. We assume that $\mathbf{Z}$ is irreducible, positive recurrent, and independent of $\mathbf{X}$. We denote by $\pi = \{\pi_i,\ i = 1, 2, \cdots\}$ the unique invariant probability measure for $\mathbf{Z}$.

The arrival process with interpoint distances $D_k = X_{Z_{k-1},Z_k}^{(k)}$, $k = 1, 2, \cdots$, is a **Markov renewal (MR)** arrival process.

The distribution of the arrival process $\mathbf{D} = \{D_1, D_2, \ldots\}$ is uniquely determined by the initial distribution $\mathbf{a} = \{a_i\}$, the transition matrix $\mathbf{A} = \{a_{ij}\}$, and the set of distribution functions $\mathbf{F} = \{F_{ij}\}$. We denote this triple by $[\mathbf{a}, \mathbf{A}, \mathbf{F}]$. The Markov chain $\mathbf{Z}$ is called the underlying Markov chain. The corresponding semi-Markov kernel is $\mathbf{A}(t) = \{a_{ij}F_{ij}(t)\}$.

It is known (see Rolski (1983)) that for the initial distribution $\mathbf{a} = \boldsymbol{\pi}$, the Markov chain $(\mathbf{D}, \mathbf{Z})$ is stationary and ergodic. In this case we call the arrival process $\mathbf{D}$ an ergodic MR arrival process.

To say that in a random vector there is more dependency than in another one with the same marginal distributions, we introduce an ordering generated by supermodular functions. A function $f : R^k \to R$ is supermodular (s-m) if for any $\mathbf{x}, \mathbf{y} \in R^k$,

$$f(\mathbf{x} \vee \mathbf{y}) + f(\mathbf{x} \wedge \mathbf{y}) \geq f(\mathbf{x}) + f(\mathbf{y}),$$

for all $\mathbf{x} = (x_1, \cdots, x_k)$, $\mathbf{y} = (y_1, \cdots, y_k)$. This condition is equivalent to $\partial^2 f/\partial x_i \partial x_j \geq 0$, for all $i \neq j$ (see e.g. Marshall and Olkin (1979)).

For the random vectors $\mathbf{X}$ and $\mathbf{Y}$ with values in R^k we write $\mathbf{X} <_{s-m} \mathbf{Y}$ if $Ef(\mathbf{X}) \leq Ef(\mathbf{Y})$ for all $f : R^k \to R$ supermodular, for which the expectations exist.

The following theorem illustrates how some positive dependency and $<_{s-m}$ ordering are related (see Theorem 3.8 in Meester and Shanthikumar (1992)).

Theorem A. *Suppose that for* $\mathbf{X} = (X_1, \cdots, X_k)$ *the conditional distributions* $P(X_k \leq x_k \mid X_1 = x_1, \cdots, X_{k-1} = x_{k-1})$ *are stochastically increasing in* $(x_1, \cdots, x_{k-1})$ *for all* k *then*

$$\hat{\mathbf{X}} <_{s-m} \mathbf{X},$$

where $\hat{\mathbf{X}} = (\hat{X}_1, \cdots, \hat{X}_k)$, $\hat{X}_i$ *are i.i.d. and* $\hat{X}_i =^d X_i$ *for* $i = 1, \cdots, k$.

The total dependence in a vector gives an upper bound in $<_{s-m}$, which is expressed by the Lorentz inequality (see Tchen (1980), Theorem 5A, Rolski (1986), Lemma 5).

Theorem B. *Suppose that for* $\mathbf{X} = (X_1, \cdots, X_n)$, $X_i, i = 1, \ldots, n$ *have a common distribution function then*

$$\mathbf{X} <_{s-m} (X_1, \cdots, X_1).$$

Consider two FCFS single server queues $MR/GI/1$ with different arrival streams. Assuming that the difference of the arrival process is caused only by a change of distribution functions in $\mathbf{F}$, Rolski (1983) proved the following theorem.

Theorem C. *Suppose that we have two MR arrival processes represented by* $[\boldsymbol{\pi}, \mathbf{A}, \mathbf{F}]$ *and* $[\boldsymbol{\pi}, \mathbf{A}, \mathbf{G}]$ *respectively to a FCFS* $MR/GI/1$ *queue. If* $F_i <_{icx} G_i$ *for* $i = 1, 2, \ldots$, *and the expected values of* F_i *and* G_i *are equal then for the corresponding stationary actual waiting times we have*

$$W[\boldsymbol{\pi}, \mathbf{A}, \mathbf{F}] <_{icx} W[\boldsymbol{\pi}, \mathbf{A}, \mathbf{G}].$$

provided the traffic intensity is smaller than 1.

Note that in this case the traffic intensity in both queues is the same. The stationary arrival intensity for both queues is given by

$$\lambda^{-1} = \sum_{j=0}^{\infty} \pi_j \int_R x dF_j(x),$$

since the mean values of F_j and G_j are the same. Hence applying Little's formula we conclude from the above theorem that the corresponding mean queue lengths are also ordered.

We now address the question how changes of $\mathbf{A}$ influence the distribution of $\mathbf{D}$ and consequently the waiting time and the queue length in $MR/GI/1$ queues.

The following theorem provides some information how changes of the governing matrix $\mathbf{A}$ imply orderings for arrival processes .

Theorem D. *Suppose that for two Markov renewal arrival processes represented by* $[\mathbf{a}, \mathbf{A}, \mathbf{F}]$ *and* $[\mathbf{a}', \mathbf{A}', \mathbf{F}']$, *respectively, the corresponding Markov chains are stochastically ordered i.e.,* $\mathbf{Z} <_{st} \mathbf{Z}'$, *and* $F_i < F_i'$, $i = 1, 2, \ldots$, *where* $<$ *denotes one of the orderings* $<_{st}$, $<_{icx}$, *or* $<_{icv}$ *(notationally,* $< = <_{st}$, $< = <_{icx}$, *or* $< = <_{icv}$*), additionally assume that* $F_i' < F_{i+1}'$, $i = 1, 2, \ldots$ *then the arrival processes are ordered with respect to the corresponding multidimensional ordering, i.e.,*

$$Ef(\mathbf{D} \mid_n) \leq Ef(\mathbf{D}' \mid_n)$$

for all f *nondecreasing in the case* $< = <_{st}$, *nondecreasing convex in the case* $< = <_{icx}$, *nondecreasing concave in the case* $< = <_{icv}$, *for which the expectations exist.*

Proof. We prove the case $<=<_{st}$. The other cases follow the same line. For a nondecreasing function $f : R^n \to R$

$$\begin{aligned}\int f(\mathbf{D}\mid_n)dP &= \iint f(x^{(1)}_{z_1},\dots,x^{(n)}_{z_n})dP^{\mathbf{X}}(\mathbf{x})dP^{\mathbf{Z}}(\mathbf{z})\\ &\le \iint f(x^{(1)}_{z_1},\dots,x^{(n)}_{z_n})dP^{\mathbf{X}'}(\mathbf{x})dP^{\mathbf{Z}}(\mathbf{z})\\ &\le \int \Phi'(\mathbf{z})dP^{\mathbf{Z}}(\mathbf{z})\\ &\le \int \Phi'(\mathbf{z})dP^{\mathbf{Z}'}(\mathbf{z}) = \int f(\mathbf{D}'\mid_n)dP.\end{aligned}$$

The first inequality follows from the fact that $F_i <_{st} F_i'$, and $\mathbf{X}$ and $\mathbf{X}'$ consist of independent random variables, the second inequality is a consequence of $\mathbf{Z} <_{st} \mathbf{Z}'$ and the fact that $\Phi'(\mathbf{z})$ is nondecreasing in $\mathbf{z}$, which is clear from the assumption that $F_i' <_{st} F_{i+1}'$, $i = 1, 2, \dots$ □

Remark. Note that from the proof of the above proposition it follows that the vectors $(D_n, \dots, D_1)$ and $(D_n', \dots, D_1')$ have the same ordering as $\mathbf{D}\mid_n$ and $\mathbf{D}'\mid_n$, respectively.

Example. (continuation). In the two dimensional case in order to have $\mathbf{Z} <_{st} \mathbf{Z}'$ it is enough to have $a \ge a'$ or $b \le b'$, if we start both processes from zero or from the corresponding stationary distributions π and π'.

Theorem E. *Suppose that in two stationary ergodic $G/G/1$ FCFS queues the stationary service time processes have the same distribution, and the stationary arrival process are such that for each $n = 1, 2, \dots$,*

$$(D_n, \dots, D_1) >_{icv} (D_n', \dots, D_1')$$

then for the corresponding stationary waiting times we have

$$W <_{icx} W'.$$

Proof. From Lemma 2.12 in Rolski (1983) we have for each ϕ nondecreasing convex

$$E\phi(W) = \lim_{n\to\infty} Eg_n(D_{-1}, \dots, D_{-n}),$$

where g_n are nonincreasing convex. Of course the functions $-g_n$ are nondecreasing concave, hence from the assumption, and stationarity of $\{D_i\}$ we have

$$E(-g_n(D_{-1}, \dots, D_{-n})) \ge E(-g_n(D_{-1}', \dots, D_{-n}'))$$

which gives $E\phi(W) \le E\Phi(W')$. □

From the above theorem we obtain a comparison result for stationary queues with the same service times and different arrival processes. We assume that the traffic intensities are smaller than 1. A $MR/GI/1$ queue with a stationary, ergodic MR arrival process is a stationary, ergodic queue since (D_n, S_n) is stationary, ergodic in this case (see Rolski (1983)).

Theorem F. *Suppose that for two stationary, ergodic MR arrival processes to a stationary, ergodic, stable MR/GI/1 queue represented by* $[\boldsymbol{\pi}, \mathbf{A}, \mathbf{F}]$ *and* $[\boldsymbol{\pi}', \mathbf{A}', \mathbf{F}']$, *respectively, the corresponding Markov chains are stochastically ordered, i.e.,* $\mathbf{Z} >_{st} \mathbf{Z}'$ *(or* $\mathbf{Z} <_{st} \mathbf{Z}'$*), and* $F_i >_{icv} F_i'$, $F_{i+1} >_{icv} F_i$ $(F_{i+1} <_{icv} F_i)$, $i = 1, 2, \ldots$ *then* $W <_{icx} W'$.

Remark. If we take $\mathbf{A} = \mathbf{A}'(\boldsymbol{\pi} = \boldsymbol{\pi}')$ and equal expectations for F_i and F_i', we get Rolski's result from Theorem C., since in this case $F_i >_{icv} F_i'$ is equivalent to $F_i <_{ic} F_i'$.

Example. (continuation). The changes of $\mathbf{A}$ in the corollary above lead in this example to making b greater but at the same time the value of a smaller, or b smaller and a greater i.e. if $\mathbf{A}$ and $\mathbf{A}'$ are stochastically ordered then $a \geq a'$ and $b \leq b'$, or $a \leq a'$ and $b \geq b'$. This causes inevitably a change of the stationary arrival process such that the one dimensional marginal distribution for this process can be quite different.

We give now two types of generalizations, first we generalize our results to the case when the holding times in the MR arrival process depend on both the actual and the previous state value, secondly we relax the strong stochastic ordering $\mathbf{Z} <_{st} \mathbf{Z}'$ to $\mathbf{Z} <_{ic} \mathbf{Z}'$ in the comparison results.

Consider a sequence of arrays of independent, positive random variables $\mathbf{X} = \{X_{ij}^{(n)},\ i, j, n = 1, 2, \ldots\}$. For each fixed i, j the sequence $\{X_{ij}^{(n)},\ n = 1, 2, \ldots\}$ is assumed to be i.i.d. with a common distribution F_{ij} . Now the arrival process with interpoint distances $D_n = X_{Z_{n-1}, Z_n}^{(n)}$, $n = 1, 2, \ldots$, is a general MR arrival process with the corresponding semi-Markov kernel $\mathbf{A}(t) = \{a_{ij} F_{ij}(t)\}$. Let $\mathbf{F} = \{F_{ij}\}$.

In order to have $\mathbf{Z} <_{ic} \mathbf{Z}'$ instead of $\mathbf{Z} <_{st} \mathbf{Z}'$ we assume some monotonicity properties for distributions in $\mathbf{F}$. We need concepts of convex increasingnees in a stochastic sense.

Let $\{X(\theta), \theta \in \Theta\}$ be a family of random variables parameterized by $\theta \in \Theta$, where Θ is a convex subset of R^n, ordered by the coordinatewise ordering. We recall the concepts of strong stochastic convexity and concavity from Shanthikumar and Yao (1991).

The family $\{X(\theta)\}$ is **strongly stochastically convex** (SSCX) if for any $\theta, \eta \in \Theta$ and any $\alpha \in [0, 1]$, there exist on a common probability space $\hat{X}(\theta) =^d X(\theta)$, $\hat{X}(\eta) =^d X(\eta)$, and $\hat{X}(\alpha\theta + (1-\alpha)\eta) =^d X(\alpha\theta + (1-\alpha)\eta)$, such that

$$\hat{X}(\alpha\theta + (1-\alpha)\eta) \leq \alpha\hat{X}(\theta) + (1-\alpha)\hat{X}(\eta) \quad (a.s.)$$

The property **strong stochastic concavity** (SSCV) is analogous to SSCX with the inequality above reversed. If, in addition, $\{X(\theta)\}$ is strongly stochastically increasing, then it satisfies strong stochastic increasing convexity or concavity, (SSICX, SSICV). Similarly we introduce decreasing families SSDCX, SSDCV, when these families are stochastically decreasing.

Theorem G. *Suppose that for two Markov renewal arrival processes represented by* $[\mathbf{a}, \mathbf{A}, \mathbf{F}]$ *and* $[\mathbf{a}', \mathbf{A}', \mathbf{F}']$, *respectively, the corresponding Markov chains are convexly (concavely) ordered, i.e.,* $\mathbf{Z} <_{icx} \mathbf{Z}'$ $(\mathbf{Z} <_{icv} \mathbf{Z}')$, *and* $F_{ij} <_{icx} F_{ij}'$, $i, j = 1, 2, \ldots$, $(F_{ij} <_{icv} F_{ij}')$, *additionally assume that for each n the family* $\{X_{ij}^{'(n)}\}$ *is SSICX (SSICV) in* $\theta = (i, j)$ *then* $\mathbf{D}\mid_n <_{icx} (<_{icv}) \mathbf{D}'\mid_n$.

Proof. For a nondecreasing convex function $f : R^n \to R$

$$\begin{aligned}\int f(\mathbf{D}\mid_n)dP &= \iint f(x^{(1)}_{z_0,z_1},\ldots,x^{(n)}_{z_{n-1},z_n})dP^{\mathbf{X}}(\mathbf{x})dP^{\mathbf{Z}}(\mathbf{z})\\ &\leq \iint f(x^{(1)}_{z_0,z_1},\ldots,x^{(n)}_{z_{n-1},z_n})dP^{\mathbf{X}'}(\mathbf{x})dP^{\mathbf{Z}}(\mathbf{z})\\ &= \int \Phi'(\mathbf{z})dP^{\mathbf{Z}}(\mathbf{z})\\ &\leq \int \Phi'(\mathbf{z})dP^{\mathbf{Z}'}(\mathbf{z}) = \int f(\mathbf{D}'\mid_n)dP.\end{aligned}$$

The first inequality follows from the fact that $F_{ij} <_{ic} F'_{ij}$, and $\mathbf{X}$ and $\mathbf{X}'$ consist of independent random variables, the second inequality is a consequence of $\mathbf{Z} <_{icx} \mathbf{Z}'$ and the fact that $\Phi'(\mathbf{z})$ is nondecreasing and convex in $\mathbf{z}$, which a consequence of the assumption that $\{X_{ij}^{'(n)}\}$ is SSICX in $\theta = (i,j)$ (see Theorem 2.11 in Shanthikumar and Yao (1991)). The case of $<_{icv}$ ordering is similar. □

Remark. Note that from the proof of the above proposition it follows that also for the vectors $(D_n,\ldots,D_1)$ and $(D'_n,\ldots,D'_1)$ we have the same ordering as for $\mathbf{D}\mid_n$ and $\mathbf{D}'\mid_n$, respectively.

From the above results we get

Corollary H. *Suppose that for two stationary, ergodic MR arrival processes to a stationary, ergodic, stable MR/GI/1 queue, represented by $[\boldsymbol{\pi},\mathbf{A},\mathbf{F}]$ and $[\boldsymbol{\pi}',\mathbf{A}',\mathbf{F}']$, respectively, the corresponding Markov chains are concavely ordered i.e., $\mathbf{Z} >_{icv} \mathbf{Z}'$, and $F_{ij} >_{icv} F'_{ij}$, $\{X_{ij}^{(n)}\}$ is SSICV in $\theta = (i,j)$ then*

$$W <_{icx} W'.$$

Remark. We can check that the family of gamma distributions $\Gamma(\alpha,\lambda)$, for $\alpha \geq 1$ is SSDCV in λ , which is a scale parameter for this family ($\lambda \geq 0$). For our queueing model in the above corollary we can take $\mathbf{F} = \{\Gamma(\alpha,\lambda_{i,j}),\ i,j = 1,2,\ldots\}$, for fixed $\alpha \geq 1$ and $\lambda_{i,j} \geq \lambda_{i',j'}$ for $(i,j) \leq (i',j')$. Some sufficient conditions for SSCX (SSCV) are given in Shanthikumar and Yao (1991)

For the class of MR arrival processes under study, the mean queue waiting time and the mean queue length in MR/GI/1 queues are finite, provided the second moment of the service time is finite. This will follow from the sufficient conditions given in Daley et al. (1992).

Theorem I. (Szekli et al. (1993)) *Consider a stationary, ergodic MR/GI/1 queue with the MR arrival process given by $[\boldsymbol{\pi},\mathbf{A},\mathbf{F}]$, i.i.d service times $\{S_n, n \geq 1\}$, and $ES_1/ED_1 < 1$. If $\mathbf{A}$ is symmetric and $ES_1^{k+1} < \infty$, $k \geq 1$ then*

$$EW^k < \infty,$$

where W denotes the stationary actual waiting time in this MR/GI/1 queue.

We now assume a special form of the transition matrix for the governing Markov chain

$$\mathbf{A}_n(p) = \begin{bmatrix} p & \frac{1-p}{n-1} & \cdots & \frac{1-p}{n-1} \\ \frac{1-p}{n-1} & p & \cdots & \frac{1-p}{n-1} \\ \vdots & \vdots & \ddots & \vdots \\ \frac{1-p}{n-1} & \frac{1-p}{n-1} & \cdots & p \end{bmatrix}, \tag{2}$$

where $n \geq 2$, $p \in (0,1)$.
For the MR arrival process represented by $[\boldsymbol{\pi}, \mathbf{A}_n(p), \mathbf{F}]$, the one-dimensional marginal distribution for the stationary interarrival time, $P(D_k \leq t)_p$, $k = 1, 2, \ldots$ is independent of p. Indeed, from Disney and Kiessler (1987) and in view of the equality $\boldsymbol{\pi} = \frac{1}{n}\mathbf{e}^T$,

$$P(D_k \leq t)_p = \boldsymbol{\pi}\mathbf{A}(t)\mathbf{e} = \frac{1}{n}\sum_{i=1}^{n} F_i(t).$$

The matrix $\mathbf{A}_n(p)$ has more desirable properties, it is symmetric, and for $p \geq \frac{1}{n}$ it is stochastically monotone . These properties imply that the corresponding ergodic arrival process $\mathbf{D}_p$ is reversible.

Lemma J. *The MR arrival process $\mathbf{D}_{\frac{1}{n}}$, represented by $[\pi, \mathbf{A}_n(\frac{1}{n}), \mathbf{F}]$ is a non-delayed renewal process.*

Proof. It is known (see Disney and Kiessler (1987)) that

$$P(D_1 \leq t_1, \ldots, D_m \leq t_m) = \boldsymbol{\pi}\mathbf{A}(t_1)\cdots\mathbf{A}(t_m)\mathbf{e},$$

for the corresponding semi-Markov kernel $\mathbf{A}(t)$, where $\mathbf{e}$ is the column vector of 1's. Now since $\mathbf{A}(t)\mathbf{e} = (\frac{1}{n}\sum_{j=1}^{n} F_j(t))\mathbf{e}$, and $\boldsymbol{\pi} = \frac{1}{n}\mathbf{e}^T$, we have

$$P(D_1 \leq t_1, \ldots, D_m \leq t_m) = (\frac{1}{n}\sum_{j=1}^{n} F_j(t_1))\cdots(\frac{1}{n}\sum_{j=1}^{n} F_j(t_m)),$$

for all $t_m \in R_+$. This implies that $\mathbf{D}_{\frac{1}{n}}$ is renewal with the interrenewal distribution function $\frac{1}{n}\sum_{j=1}^{n} F_j(t)$. □

For a stationary ergodic MR/GI/1 queue with the MR arrival process given by $[\boldsymbol{\pi}, \mathbf{A}_n(p), \mathbf{F}]$, i.i.d sequence of service times $\{S_n\}$, and $ES_1/ED_1 < 1$, denote by W_p the stationary actual waiting time in this queue. Note that the traffic intensity $\rho = ES_1/ED_1$ does not depend on p in this class of queues.

Theorem K. *Consider two stationary ergodic MR/GI/1 queues ($\rho < 1$) with the interarrival processes represented by $[\boldsymbol{\pi}, \mathbf{A}_n(\frac{1}{n}), \mathbf{F}]$, $[\boldsymbol{\pi}, \mathbf{A}_n(p), \mathbf{F}]$ ($p \in (\frac{1}{n}, 1)$), respectively. If $F_j <_{st} F_{j+1}$, $j = 1, \cdots, n-1$, then*

$$W_{\frac{1}{n}} <_{icx} W_p.$$

The proof of Theorem K. is a consequence of the following two lemmas. The first one is stated in a different form (formulated for $<_{idcx}$ ordering) as Theorem 3.12 in Meester and Shanthikumar (1992).

Lemma L. *For stationary ergodic MR arrival processes* $[\boldsymbol{\pi}, \mathbf{A}_n(p), \mathbf{F}]$ *with* $p \geq \frac{1}{n}$ *and* $F_j <_{st} F_{j+1}$, $j = 1, 2, \cdots$, *we have for* $k = 1, 2, \cdots$,

$$\mathbf{D}_{\frac{1}{n}}|_k <_{s-m} \mathbf{D}_p|_k,$$

where $\mathbf{D}_p|_k$ *denotes the vector of the first* k *coordinates of* $\mathbf{D}_p$.

Proof. Denote by $\mathbf{Z}_p$ the stationary underlying Markov chain corresponding to $\mathbf{A}_n(p)$. For $p > \frac{1}{n}$, $\mathbf{A}_n(p)$ is stochastically monotone, hence the assumptions of Theorem A. are fulfilled for the vector $\mathbf{Z}_p|_k$. The corresponding "independent" version is represented by $\mathbf{Z}_{\frac{1}{n}}|_k$ (see Lemma J.) , hence we have $\mathbf{Z}_{\frac{1}{n}}|_k <_{s-m} \mathbf{Z}_p|_k$ for all k.
Consider a vector $D(\mathbf{x}, \mathbf{z}) = (x^{(1)}_{z_1}, \cdots, x^{(k)}_{z_k})$, where

$$\mathbf{x} = \begin{bmatrix} x_1^{(1)} & \cdots & x_1^{(k)} \\ x_2^{(1)} & \cdots & x_2^{(k)} \\ \vdots & \vdots & \vdots \\ x_n^{(1)} & \cdots & x_n^{(k)} \end{bmatrix}, \qquad \mathbf{z} = (z_1, \cdots, z_k).$$

Denote by $P^{\mathbf{X}}$ the distribution of $\mathbf{X}$. Since we can have $\mathbf{X}_i^{(k)} \leq \mathbf{X}_{i+1}^{(k)}$ a.s. on some Ω (for all i and k), from the assumption that $F_i <_{st} F_{i+1}$ (see e.g. Stoyan (1983), Proposition 1.10.4, and (1.10.5)), $D(\mathbf{x}, \mathbf{z})$ is increasing a.s. $P^{\mathbf{X}}$ in $\mathbf{z}$. This implies that $f(D(\mathbf{x}, \mathbf{z}))$ is supermodular in $\mathbf{z}$ (a.s. $P^{\mathbf{X}}$) for each supermodular function f. Denoting by $E_{\mathbf{X}}$ the expectation with respect to $P^{\mathbf{X}}$, we have

$$\begin{aligned} Ef(\mathbf{D}_{\frac{1}{n}}|_k) &= E_{\mathbf{X}} E_{\mathbf{Z}_{\frac{1}{n}}|_k} f(D(\mathbf{X}, \mathbf{Z}_{\frac{1}{n}}|_k)) \\ &\leq E_{\mathbf{X}} E_{\mathbf{Z}_p|_k} f(D(\mathbf{X}, \mathbf{Z}_p|_k)) = Ef(\mathbf{D}_p|_k), \end{aligned}$$

i.e., $\mathbf{D}_{\frac{1}{n}}|_k <_{s-m} \mathbf{D}_p|_k$. □

The following lemma is a useful modification of Lemma 2.12 of Rolski (1983).

Lemma M. *If* W *is a stationary actual waiting time in a stable G/GI/1 FCFS queue with a stationary reversible interarrival stream* $\{D_k\}$, *then for each increasing and convex function* φ,

$$E\varphi(W) = \lim_{k \to \infty} Eg_k(D_1, \cdots, D_k),$$

for a sequence of supermodular functions g_k.

Proof. We have (see e.g., Loynes (1962)) the following representation

$$W =^d \sup(0, S_{-1} - D_{-1}, S_{-1} + S_{-2} - D_{-1} - D_{-2}, \cdots),$$

where $\{S_i\}$ is the sequence of service times. Since $\{D_i\}$ is reversible, we have

$$W =^d \sup(0, S_1 - D_1, S_1 + S_2 - D_1 - D_2, \cdots)$$

and

$$\varphi(W) = \varphi(\sup(0, S_1 - D_1, S_1 + S_2 - D_1 - D_2, \cdots)).$$

Consider a function $f_k : R^{2k} \to R$ such that $f_k(\mathbf{x}, \mathbf{s}) = \max(0, s_1 - x_1, s_1 + s_2 - x_1 - x_2, \cdots, s_1 + \cdots + s_k - x_1 - \cdots - x_k)$, for $\mathbf{s} = (s_1, \cdots, s_k)$. Rolski (1986) proved that for each $\mathbf{s}$, $f_k(-\mathbf{x}, \mathbf{s})$ is supermodular in $\mathbf{x}$. From this we have $f_k(\mathbf{x}, \mathbf{s})$ is also supermodular in $\mathbf{x}$. Hence

$$\varphi(W) = \lim_{k\to\infty} \varphi \circ f_k((D_1, \cdots, D_k), (S_1, \cdots, S_k)).$$

and

$$\begin{aligned} E\varphi(W) &= E_{\mathbf{S}} E_{\mathbf{D}} \varphi(W) \\ &= E_{\mathbf{S}} E_{\mathbf{D}} \lim_{k\to\infty} \varphi \circ f_k((D_1, \cdots, D_k), (S_1, \cdots, S_k)) \\ &= \lim_{k\to\infty} E_{\mathbf{S}|_k} E_{\mathbf{D}|_k} \varphi \circ f_k((D_1, \cdots, D_k), (S_1, \cdots, S_k)), \end{aligned}$$

by the monotone convergence theorem. Let $P^{\mathbf{D}|_k}$ and $P^{\mathbf{S}|_k}$ be the distribution functions of $(D_1, \cdots, D_k)$ and $(S_1, \cdots, S_k)$ respectively. Then

$$\begin{aligned} & E_{\mathbf{S}|_k} E_{\mathbf{D}|_k} \varphi \circ f_k((D_1, \cdots, D_k), (S_1, \cdots, S_k)) \\ &= \int_{R_+^k} \int_{R_+^k} \varphi \circ f_k((d_1, \cdots, d_k), (s_1, \cdots, s_k))\, dP^{\mathbf{D}|_k}(\mathbf{d}) dP^{\mathbf{S}|_k}(\mathbf{s}) \\ &= E g_k(\mathbf{D}|_k), \end{aligned}$$

where $g_k(\mathbf{D}|_k) = \int_{R_+^k} \varphi \circ f_k((d_1, \cdots, d_k), (s_1, \cdots, s_k))\, dP^{\mathbf{S}|_k}(\mathbf{s})$.
Therefore, $E\varphi(W) = \lim_{k\to\infty} E g_k(\mathbf{D}|_k)$. Since mixtures of supermodular functions are supermodular and for each φ increasing convex, $\varphi \circ f_k$ is a supermodular function, g_k's are supermodular. □

Proof of Theorem K.. From Lemma M.

$$E\varphi(W_{\frac{1}{n}}) = \lim_{k\to\infty} E g_k(\mathbf{D}_{\frac{1}{n}}|_k),$$

and

$$E\varphi(W_p) = \lim_{k\to\infty} E g_k(\mathbf{D}_{\mathbf{p}}|_k),$$

since $\mathbf{D}_p$ is reversible for $p \geq \frac{1}{n}$. Now from Lemma L.

$$E g_k(\mathbf{D}_{\frac{1}{n}}|_k) \leq E g_k(\mathbf{D}_p|_k),$$

since g_k's are supermodular. By the definition of $<_{icx}$ the proof is complete. □

We obtain an upper bound for W_p in a similar way, using the Lorentz inequality (see Theorem B.).

Theorem N. *For stationary ergodic MR arrival processes* $[\boldsymbol{\pi}, \mathbf{A}_n(p), \mathbf{F}]$ *with* $p \geq \frac{1}{n}$, *and* $F_j <_{st} F_{j+1}$, $j = 1, 2, \cdots, n-1$, *we have*

$$\mathbf{D}_p|_k <_{s-m} \mathbf{D}^u|_k,$$

where $\mathbf{D}^u = (D_1^u, D_2^u, \cdots)$ *corresponds to* $[\boldsymbol{\pi}, \mathbf{A}_n(1), \mathbf{F}]$.

The following theorem is a consequence of Lemma M..

Theorem O. *Consider two stationary ergodic MR/GI/1 queues ($\rho < 1$) with the interarrival processes represented by* $[\boldsymbol{\pi}, \mathbf{A}_n(p), \mathbf{F}]$, $[\boldsymbol{\pi}, \mathbf{A}_n(1), \mathbf{F}]$ *($p \in (\frac{1}{n}, 1)$), respectively. If $F_j <_{st} F_{j+1}$, $j = 1, \cdots, n-1$, then*

$$W_p <_{icx} W_1.$$

Some lower bounds for MR/GI/1 queues with $[\boldsymbol{\pi}, \mathbf{A}_n(p), \mathbf{F}]$ arrival process can be given in terms of other GI/GI/1 systems, if we assume in addition that F_j's are gamma, Weibull, or DFR. These bounds are slightly lower than that given by $W_{\frac{1}{n}}$ but they seem to be more tractable, and thus of interest.

Theorem P. (Szekli et al. (1993b)) *For a stationary ergodic MR/GI/1 queue with the arrival process represented by* $[\boldsymbol{\pi}, \mathbf{A}_n(p), \mathbf{F}]$ *with $F_j <_{st} F_{j+1}$, if*

(i) *F_j's are gamma with the same shape parameter, or*

(ii) *F_j's are Weibull with the same shape parameter, or*

(iii) *F_j's are DFR,*

then we have

$$W^{\ell} <_{icx} W_p,$$

where W^{ℓ} is a stationary waiting time in a GI/GI/1 queue with the same service times and the same mean interarrival times as in the MR/GI/1 queue, but interarrivals to this queue are gamma with the same shape parameter as F_j's in the case (i), Weibull with the same shape parameter as F_j's in the case (ii) and exponential in the case (iii).

Chapter 3

Dependence

One of the fundamental problems of positive dependence has been to obtain conditions on a multivariate vector $\mathbf{X} = (X_1, \ldots, X_n)$ such that the condition

$$P(X_1 > x_1, \ldots, X_n > x_n) \geq \prod_{i=1}^{n} P(X_i > x_i),$$

or conditions similar to this hold, for all real x_i. One of the best known notions, which implies this type of inequalities is **association**. The vector $\mathbf{X}$ is associated if $Cov(f(\mathbf{X}), g(\mathbf{X})) \geq 0$, for every pair of increasing functions $f, g : R^n \to R$. Implicit in a conclusion that a set of random variables is associated is a wealth of inequalities, often of direct use in various statistical problems. There are two almost independent parts of literature on the subject of associated random variables. One developed from the work of Esary, Proschan and Walkup (1967), and is oriented towards reliability theory and statistics. The other, developed from the works of Harris (1960) and Fortuin, Kastelyn and Ginibre (1971), is oriented towards percolation theory and statistical mechanics. In the statistical mechanics literature associated random variables are said to satisfy the FKG inequalities. Although the recognition that association is also useful in the study of approximate independence, seems to have first occurred in Lebowitz (1972), the main contributions to the study of independence and limit theorems for associated random variables include works of Newman and co-authors, which is reviewed by Newman (1984).

Since association represents a strong positive dependence, weaker concepts have been considered in the literature. Many of these can be viewed as variation of the classes, from which f, g are chosen and then

$$Cov(f(\mathbf{X}), g(\mathbf{X})) \geq 0$$

imposed. A general theory of some positive dependence notions includes the work by Shaked (1982).

Notions of negative dependence have received relative very little attention in the literature. The main motivation for definitions in this direction is to try to formulate the intuitive requirement that if a set of negatively dependent random variables is split into two subsets in some manner then one subset will tend to be "large" when the other subset "small", and vice versa.

Important contributions directed to the objective of characterizing the notions of dependence include works of Ahmed, Leon and Proschan (1981), Block , Savits and Shaked (1982), Joag-Dev and Proschan (1983), Shaked and Shanthikumar (1989), among others.

3.1 Association

For real random variables $X_1, \ldots, X_n$, we write $\mathbf{X} = (X_1, \ldots, X_n)$ to denote random vector with values in R^n.

We say that a random vector $\mathbf{X}$ is associated if

$$Cov(f(\mathbf{X}), g(\mathbf{X})) \geq 0, \tag{1}$$

for all nondecreasing real functions f, g, for which the covariance exists.

Association has the following properties (see Esary et al. (1967))

P1 Any subset of associated random variables is associated;

P2 If two sets of associated random variables are independent of one another then their union is a set of associated random variables;

P3 The set consisting of a single random variable is associated;

P4 Non-decreasing functions of associated random variables are associated;

P5 If $\mathbf{X}^{(k)}$ are associated, for each k, and $\mathbf{X}^{(k)} \to \mathbf{X}$ in distribution then $\mathbf{X}$ is associated.

Association of random variables may be established by taking in (1) nondecreasing test functions f, g, which are binary, or bounded and continuous.

The following identity will be useful in many situations.

Lemma A. *For two random variables X, Y*

$$Cov(X,Y) = \int_{-\infty}^{\infty}\int_{-\infty}^{\infty} Cov(I_{(X>s)}, I_{(Y>t)})dsdt. \tag{2}$$

Proof. Usually this identity is proved from integration by parts. A simple proof is possible also from Fubini's theorem.
Suppose first that $X, Y \geq 0$. Since

$$Cov(I_{(X>s)}, I_{(Y>s)}) = EI_{(X>s)}I_{(Y>t)} - EI_{(X>s)}EI_{(Y>t)},$$

we have to find

$$\int_{-\infty}^{\infty}\int_{-\infty}^{\infty} EI_{(X>s)}EI_{(Y>t)}dsdt,$$

and

$$\int_{-\infty}^{\infty}\int_{-\infty}^{\infty} E(I_{(X>s)}I_{(Y>t)})dsdt.$$

We have

$$\int_{-\infty}^{\infty}\int_{-\infty}^{\infty} EI_{(X>s)}EI_{(Y>t)}dsdt = \int_{0}^{\infty} P(X>s)ds\int_{0}^{\infty} P(Y>t)dt = EXEX.$$

$$\int_{-\infty}^{\infty}\int_{-\infty}^{\infty} E(I_{(X>s)}I_{(Y>t)})dsdt = \int_{-\infty}^{\infty}\int_{-\infty}^{\infty}\int_{\Omega} I_{(X>s)}I_{(Y>t)}dPdsdt =$$

$$\text{from Fubini's th.} = \int_{\Omega}\int_{-\infty}^{\infty}\int_{-\infty}^{\infty} I_{(X>s)}I_{(Y>t)}dsdtdP =$$

$$\int_{\Omega}\int_{-\infty}^{\infty} I_{(Y>t)}dt\int_{-\infty}^{\infty} I_{(X>s)}dsdP = \int_{\Omega} YXdP = EXY.$$

Hence $\int_{-\infty}^{\infty}\int_{-\infty}^{\infty} Cov(I_{(X>s)}I_{(Y>t)})dsdt = EXY - EXEY = Cov(X,Y)$.

This completes the proof, for $X, Y \geq 0$.
Suppose now that X, Y are bounded below by some $\Delta \in R$. Then $X-\Delta \geq 0$, $Y-\Delta \geq 0$. Applying the first part of the proof to them we again obtain the identity. Letting $\Delta \to -\infty$ completes the proof for general random variables. □

Corollary B.

$$Cov(f(\mathbf{X}), g(\mathbf{X})) = \int_{-\infty}^{\infty}\int_{-\infty}^{\infty} Cov(I_{(f(\mathbf{X})>s)}I_{(g(\mathbf{X})>t)})dsdt,$$

for f, g, for which the integrals exist.

Since $I_{(f(\mathbf{X})>s)}$ and $I_{(g(\mathbf{X})>t)}$ are binary (0 or 1 valued) random variables, we have from the above

Corollary C. *If (1) is fulfilled for all binary nondecreasing functions f, g then $\mathbf{X}$ is associated.*

Note that the covariance on the right-hand side of (2) can be rewritten as

$$Cov(I_{(X>s)}I_{(Y>t)}) = P(X>s, Y>t) - P(X>s)P(Y>t).$$

We denote the above value by $H_{X,Y}(s,t)$, so in the new notation,

$$Cov(X,Y) = \int_{-\infty}^{\infty}\int_{-\infty}^{\infty} H_{X,Y}(s,t)dsdt. \tag{3}$$

Such a representation we use later to characterize independence via uncorelatedness.

If X, Y are associated then it is immediate to see that in this case $H_{X,Y}(s,t) \geq 0$, for each s,t. If additionally we know that X, Y are uncorrelated then from (3) we get $H_{X,Y}(s,t) = 0$, for all s, t. Hence we have

Corollary D. *If X and Y are associated then they are independent if and only if $Cov(X,Y) \geq 0$.*

The condition $H_{X,Y}(s,t) \geq 0$ does not imply association. We say that X, Y are positively quadrant dependent (PQD) if $H_{X,Y}(s,t) \geq 0$, for each $s, t \in R$. Similarly we define negative quadrant dependent (NQD) random variables.

We reformulate the last corollary as follows

Theorem E. (Lehmann(1966)) *If X and Y are PQD or NQD then they are independent if and only if $Cov(X,Y) = 0$.*

Some generalizations of this theorem will be presented later.

Examples.
1. Let $X_1, \ldots, X_n$ be independent. If $S_n = \sum_{i=1}^n X_i$ then $(S_1, \ldots, S_n)$ are associated.
2. $X_{(1)}, \ldots, X_{(n)}$, the order statistics in a sample $X_1, \ldots, X_n$, are associated.
3. (Marshall and Olkin (1966)) Consider the multivariate distribution

$$F(x_1, \ldots, x_n) = 1 - \exp[-\sum_1^n \lambda_i x_i - \sum_{i<j} \lambda_{ij} \max(x_i, x_j) - \ldots - \max(x_1, \ldots, x_n)].$$

If $\mathbf{X}$ is distributed according to F then $\mathbf{X}$ is associated. It follows from the representation

$$X_j = \min(Y_i; i \in A_j), A_j \subseteq \{1, \ldots, n\},$$

for independent exponential random variables $Y_1, \ldots, Y_m$. Each X_j is a nondecreasing function of $\mathbf{Y}$.

The notion of association can be extended from R^n to $\mathbf{X}_{i=1}^n \mathbf{Z}_i$, where each $\mathbf{Z}_i$ is a linearly ordered space (with $\leq_i$) (see Ahmed et al. (1981)).

For $\mathbf{a}, \mathbf{b} \in \mathbf{X}_{i=1}^n \mathbf{Z}_i$ we define a natural coordinatewise ordering

$$\mathbf{a} \leq \mathbf{b} \text{ if } a_i \leq_i b_i, \; i = 1, \ldots, n.$$

A random vector $\mathbf{X}$ with values in $\mathbf{X}_{i=1}^n \mathbf{Z}_i$ is associated($\leq$) if (1) is fulfilled, for all f, g nondecreasing with respect to $\leq$, for which the integrals exist. The assumption of linearly ordered $\mathbf{Z}_i$ is crucial. As a consequence, the fundamental properties **P1-P5** of association carry over to association($\leq$).

In the special case:

$$\mathbf{X}_{i=1}^n \mathbf{Z}_i = \mathcal{X}^k \times \mathcal{Y}^k, \; k + l = n,$$

where $\mathcal{X}, \mathcal{Y}$ are linearly ordered spaces, we have an interesting result

Theorem F. (Jogdeo (1978)) *Suppose that a random vector $\mathbf{X}$ with values in $\mathcal{X}^k$ is associated-($\leq$), a random vector $\mathbf{Y}$ with values in $\mathcal{Y}^l$ is conditionally, given $\mathbf{X}$, associated-($\leq$), and*

$$E(f(\mathbf{Y}) \mid \mathbf{X} = \mathbf{x})$$

is non-decreasing in $\mathbf{x}$, for all non-decreasing real functions f then $(\mathbf{X}, \mathbf{Y})$ is associated-($\leq$). In particular $\mathbf{Y}$ is associated-($\leq$).

Example. (Jogdeo (1977)).
Suppose $\mathcal{X}^k = R^k, \leq = <^a$, where $\mathbf{a} <^a \mathbf{b}$ if $| a_i | \leq | b_i |, i = 1, \ldots$,k. Let $\mathbf{Z}$ be a vector with mutually independent components, each having a symmetric unimodal density. Let $\mathbf{Y} = \mathbf{Z} + \mathbf{X}$, where $\mathbf{X}$ is independent of $\mathbf{Z}$, and $(| X_1 |, \ldots, | X_n |)$ is associated. Then $(| Y_1 |, \ldots, | Y_l |)$ is associated.

Further generalizations of the concept of association are known for vectors with values in a partially ordered Polish space (see Lindqvist (1988)).

3.2 MTP_2

The abstract characterizations of association are often difficult to deal with in a concrete statistical context. Therefore a stronger notion than association but better analytically tractable may be of interest. Such a stronger notion is for example the MTP_2 property, introduced by Karlin and Rinott (1980). We assume in this subsection that a random vector $\mathbf{X}$ takes on values in a space $\mathcal{X} = \mathcal{X}_1 \times \ldots \times \mathcal{X}_n$, where each $\mathcal{X}_i$ is a totally ordered space equipped with a σ-finite measure σ_i . On $\mathcal{X}$ we consider the product measure $\sigma = \sigma_1 \times \ldots \times \sigma_n$. For simplicity, we write $d\mathbf{x}$ for $d\sigma(\mathbf{x})$, for integration. Monotonicity is always regarded with respect to the coordinatewise ordering. We assume additionally that the distribution of $\mathbf{X}$ possesses a density $\mathbf{f}$ with respect to σ .

Definition A. *A density* $\mathbf{f}$ *is* MTP_2 *(multidimensional totally positive) if*

$$\mathbf{f}(\mathbf{x})\mathbf{f}(\mathbf{y}) \leq \mathbf{f}(\mathbf{x} \wedge \mathbf{y})\mathbf{f}(\mathbf{x} \vee \mathbf{y}),$$

for $\mathbf{x}, \mathbf{y} \in \mathcal{X}$ *, where*
$\mathbf{x} \wedge \mathbf{y} = (\min(x_1, y_1), \ldots, \min(x_n, y_n))$, $\mathbf{x} \vee \mathbf{y} = (\max(x_1, y_1), \ldots, \max(x_n, y_n))$.

We say also that $\mathbf{X}$ is MTP_2 if its density $\mathbf{f}$ is MTP_2 .
The following basic technical lemma will be used to relate the notions of MTP_2, association, and stochastic ordering.

Lemma B. (Karlin (1980)) *Let* $\mathbf{f}_1, \mathbf{f}_2, \mathbf{f}_3, \mathbf{f}_4$ *be nonnegative functions on* $\mathcal{X}$ *satisfying for all* $\mathbf{x}, \mathbf{y} \in \mathcal{X}$

$$\mathbf{f}_1(\mathbf{x})\mathbf{f}_2(\mathbf{y}) \leq \mathbf{f}_3(\mathbf{x} \vee \mathbf{y})\mathbf{f}_4(\mathbf{x} \wedge \mathbf{y}).$$

Then

$$\int \mathbf{f}_1(\mathbf{x})d\mathbf{x} \int \mathbf{f}_2(\mathbf{x})d\mathbf{x} \leq \int \mathbf{f}_3(\mathbf{x})d\mathbf{x} \int \mathbf{f}_4(\mathbf{x})d\mathbf{x}.$$

Proof. We first proof the inequality for one dimensional vectors, i.e. $n = 1$. Then we proceed by induction on n .

$$\int \mathbf{f}_1(\mathbf{x})d\mathbf{x} \int \mathbf{f}_2(\mathbf{x})d\mathbf{x} = \int \int \mathbf{f}_1(\mathbf{x})\mathbf{f}_2(\mathbf{y})d\mathbf{x}d\mathbf{y} =$$

$$= \int_{\mathbf{x}<\mathbf{y}} \int \mathbf{f}_1(\mathbf{x})\mathbf{f}_2(\mathbf{y})d\mathbf{x}d\mathbf{y} + \int_{\mathbf{y}<\mathbf{x}} \int \mathbf{f}_1(\mathbf{x})\mathbf{f}_2(\mathbf{y})d\mathbf{x}d\mathbf{y} + \int_{\mathbf{x}=\mathbf{y}} \int \mathbf{f}_1(\mathbf{x})\mathbf{f}_2(\mathbf{y})d\mathbf{x}d\mathbf{y} =$$

$$= \int_{\mathbf{x}<\mathbf{y}} \int (\mathbf{f}_1(\mathbf{x})\mathbf{f}_2(\mathbf{y}) + \mathbf{f}_1(\mathbf{y})\mathbf{f}_2(\mathbf{x}))d\mathbf{x}d\mathbf{y} + \int_{\mathbf{x}=\mathbf{y}} \int \mathbf{f}_1(\mathbf{x})\mathbf{f}_2(\mathbf{y})d\mathbf{x}d\mathbf{y}.$$

Since

$$\mathbf{f}_1(\mathbf{x})\mathbf{f}_2(\mathbf{x}) \leq \mathbf{f}_3(\mathbf{x})\mathbf{f}_4(\mathbf{x}),$$

for $\mathbf{x} \in \mathcal{X}$ it is enough to confirm the inequality

$$\int_{\mathbf{x}<\mathbf{y}} \int (\mathbf{f}_1(\mathbf{x})\mathbf{f}_2(\mathbf{y}) + \mathbf{f}_1(\mathbf{y})\mathbf{f}_2(\mathbf{x}))d\mathbf{x}d\mathbf{y} \leq \int_{\mathbf{x}<\mathbf{y}} \int (\mathbf{f}_3(\mathbf{x})\mathbf{f}_4(\mathbf{y}) + \mathbf{f}_3(\mathbf{y})\mathbf{f}_4(\mathbf{x}))d\mathbf{x}d\mathbf{y}.$$

For fixed $\mathbf{x} \leq \mathbf{y}$ let

$$\begin{aligned} \mathbf{f}_1(\mathbf{x})\mathbf{f}_2(\mathbf{y}) &= a, \\ \mathbf{f}_1(\mathbf{y})\mathbf{f}_2(\mathbf{x}) &= b, \\ \mathbf{f}_3(\mathbf{x})\mathbf{f}_4(\mathbf{y}) &= c, \\ \mathbf{f}_3(\mathbf{y})\mathbf{f}_4(\mathbf{x}) &= d. \end{aligned}$$

From the assumption we get $a, b \leq d$, and

$$\mathbf{f}_1(\mathbf{x})\mathbf{f}_2(\mathbf{x}) \leq \mathbf{f}_3(\mathbf{x})\mathbf{f}_4(\mathbf{x}),$$

$$\mathbf{f}_1(\mathbf{y})\mathbf{f}_2(\mathbf{y}) \leq \mathbf{f}_3(\mathbf{y})\mathbf{f}_4(\mathbf{y}).$$

Multiplying this side by side we have $ab \leq cd$. Now from the identity

$$c + d - (a + b) = [(d - a)(d - b) + (cd - ab)],$$

we immediately obtain $a + b \leq c + d$, which completes the proof for $n = 1$. Consider now the marginals

$$\phi_i(\mathbf{x}) = \int_{\mathcal{X}_n} \mathbf{f}_i(\mathbf{x}, x)d\sigma_n(x), i = 1, 2, 3, 4,$$

where $\mathbf{x} \in \mathbf{X}_{i=1}^{n-1}\mathcal{X}_i, x \in \mathcal{X}_n$. For these $n-1$ dimensional functions we have

$$\phi_1(\mathbf{x})\phi_2(\mathbf{y}) \leq \phi_3(\mathbf{x} \vee \mathbf{y})\phi_4(\mathbf{x} \wedge \mathbf{y}).$$

Indeed, it is enough to confirm the inequality

$$\int_{\mathbf{x}<\mathbf{y}} \int (\mathbf{f}_1(\mathbf{x}, x)\mathbf{f}_2(\mathbf{y}, y) + \mathbf{f}_1(\mathbf{x}, y)\mathbf{f}_2(\mathbf{y}, x))d\sigma_n(x)d\sigma_n(y) \leq$$

$$\leq \int_{\mathbf{x}<\mathbf{y}} \int (\mathbf{f}_3(\mathbf{x} \vee \mathbf{y}, x)\mathbf{f}_4(\mathbf{x} \wedge \mathbf{y}, y) + \mathbf{f}_3(\mathbf{x} \vee \mathbf{y}, y)\mathbf{f}_4(\mathbf{x} \wedge \mathbf{y}, x))d\sigma_n(x)d\sigma_n(y).$$

This obtains as before by setting

$$\begin{aligned} \mathbf{f}_1(\mathbf{x}, x)\mathbf{f}_2(\mathbf{y}, y) &= a, \\ \mathbf{f}_1(\mathbf{x}, y)\mathbf{f}_2(\mathbf{y}, x) &= b, \\ \mathbf{f}_3(\mathbf{x} \vee \mathbf{y}, x)\mathbf{f}_4(\mathbf{x} \wedge \mathbf{y}, y) &= c, \\ \mathbf{f}_3(\mathbf{x} \vee \mathbf{y}, y)\mathbf{f}_4(\mathbf{x} \wedge \mathbf{y}, x) &= d. \end{aligned}$$

By the induction hypotheses

$$\int \phi_1(\mathbf{x})d\mathbf{x} \int \phi_2(\mathbf{x})d\mathbf{x} \leq \int \phi_3(\mathbf{x})d\mathbf{x} \int \phi_4(\mathbf{x})d\mathbf{x},$$

which at hand gives the required inequality

$$\int \mathbf{f}_1(\mathbf{x})d\mathbf{x} \int \mathbf{f}_2(\mathbf{x})d\mathbf{x} \leq \int \mathbf{f}_3(\mathbf{x})d\mathbf{x} \int \mathbf{f}_4(\mathbf{x})d\mathbf{x},$$

where in the first expression $d\mathbf{x}$ denotes integration with respect to $n-1$ dimensional product measure, while in the second expression, with respect to n dimensional product measure. The proof is complete. □

Before we relate MTP_2 to association, we introduce TP_2 ordering.

Definition C. *For two nonnegative functions on $\mathbf{f}_1, \mathbf{f}_2$ defined on $\mathcal{X}$ we write*

$$\mathbf{f}_1 <_{TP_2} \mathbf{f}_2 \quad \textit{if} \quad \mathbf{f}_1(\mathbf{x})\mathbf{f}_2(\mathbf{y}) \leq \mathbf{f}_1(\mathbf{x} \wedge \mathbf{y})\mathbf{f}_2(\mathbf{x} \vee \mathbf{y}).$$

We say that two random vectors $\mathbf{X}_1, \mathbf{X}_2$ with values in $\mathcal{X}$ are ordered in TP_2 ordering ($\mathbf{X}_1 <_{TP_2} \mathbf{X}_2$) if their densities fulfill $\mathbf{f}_1 <_{TP_2} \mathbf{f}_2$.

Remark. Note that $\mathbf{f} <_{TP_2} \mathbf{f}$ means that $\mathbf{f}$ is MTP_2 .

Theorem D.

If $\mathbf{X}_1 <_{TP_2} \mathbf{X}_2$ *then* $\mathbf{X}_1 <_{st} \mathbf{X}_2$.

Proof. Suppose that ϕ is a nondecreasing, nonnegative function $\phi : \mathcal{X} \to \mathbf{R}$. From the basic technical lemma by substituting

$$\mathbf{f}_1^* = \mathbf{f}_1\phi, \mathbf{f}_2^* = \mathbf{f}_2, \mathbf{f}_3^* = \mathbf{f}_2\phi, \mathbf{f}_4^* = \mathbf{f}_1,$$

we obtain

$$\int \phi(\mathbf{x})\mathbf{f}_1(\mathbf{x})d\mathbf{x} \leq \int \phi(\mathbf{x})\mathbf{f}_2(\mathbf{x})d\mathbf{x}.$$

The assumption that ϕ is nonnegative can be relaxed by subtracting a constant, truncating first if necessary and using a limiting argument. The above inequality then implies $\mathbf{X}_1 <_{st} \mathbf{X}_2$. □

Theorem E. *If* $\mathbf{X}$ *is* MTP_2 *then* $\mathbf{X}$ *is associated.*

Proof. Assume that $\psi(\mathbf{x})$ is a positive nondecreasing function and define

$$\mathbf{f}_1(\mathbf{x}) = \mathbf{f}(\mathbf{x}), \mathbf{f}_2(\mathbf{x}) = \psi(\mathbf{x})\mathbf{f}(\mathbf{x}) / \int \psi(\mathbf{x})\mathbf{f}(\mathbf{x})\ d\mathbf{x}.$$

Since $\mathbf{f}$ is MTP_2 and ψ is non-decreasing we have

$$\mathbf{f}(\mathbf{x})\mathbf{f}(\mathbf{y})\psi(\mathbf{y}) \leq \mathbf{f}(\mathbf{x} \wedge \mathbf{y})\mathbf{f}(\mathbf{x} \vee \mathbf{y})\psi(\mathbf{x} \vee \mathbf{y}),$$

for $\mathbf{x}, \mathbf{y} \in \mathcal{X}$, hence $\mathbf{f}_1 <_{TP_2} \mathbf{f}_2$. From the previous theorem it follows now

$$\int \psi(\mathbf{x})\mathbf{f}(\mathbf{x})d\mathbf{x} \int \phi(\mathbf{x})\mathbf{f}(\mathbf{x})d\mathbf{x} \leq \int \psi(\mathbf{x})\psi(\mathbf{x})\mathbf{f}(\mathbf{x})d\mathbf{x},$$

for any nondecreasing ϕ. The assumption that ψ is positive can be omitted as in the previous theorem, so from the above we get for all non-decreasing ϕ, ψ

$$Cov(\phi(\mathbf{X}), \psi(\mathbf{X})) \geq 0,$$

which completes the proof. □

It is clear that we can take in the last theorem ϕ, ψ both nonincreasing.

The class MTP_2 possesses many nice properties, for example (see Karlin and Rinott (1980))

1. If $\mathbf{f}, \mathbf{g}$ are MTP_2 then $\mathbf{fg}$ is MTP_2 .
2. A vector $\mathbf{X}$ of independent random variables is MTP_2.
3. The marginals of $\mathbf{f}$, which is MTP_2, are also MTP_2.
4. If $\mathbf{f}(\mathbf{x}, \mathbf{y})$ is MTP_2 on $\mathcal{X} \times \mathcal{Y}$, $\mathbf{g}(\mathbf{y}, \mathbf{z})$ is MTP_2 on $\mathcal{Y} \times \mathcal{Z}$ then

$$\mathbf{h}(\mathbf{x}, \mathbf{z}) = \int \mathbf{f}(\mathbf{x}, \mathbf{y})\mathbf{g}(\mathbf{y}, \mathbf{z})d\mathbf{y}$$

is MTP_2 on $\mathcal{X} \times \mathcal{Z}$, where the spaces $\mathcal{Y}, \mathcal{Z}$ have the same structure as $\mathcal{X}$.

5. If $\mathbf{f}$ is MTP_2 then $\mathbf{f}(\phi_1(x_1), \ldots, \phi_n(x_n))$ is MTP_2 , where ϕ_i are nondecreasing.

PROBLEMS AND REMARKS

A. If $\mathbf{X}$ is a vector of i.i.d. random variables with PF_2 density, $\mathbf{Y}$ is MTP_2 on $\mathbf{R}^n$, and is independent of $\mathbf{X}$ then $\mathbf{Z} = \mathbf{X} + \mathbf{Y}$ is MTP_2 on $\mathbf{R}^n$. This is immediate from the properties 1-4 above, and the following formula, which gives a joint density of $\mathbf{Z}$

$$\mathbf{f_Z}(z_1, \ldots, z_n) = \int (\prod_{i=1}^{n} \mathbf{f}_{\mathbf{X}_i}(z_i - y_i))\mathbf{f_Y}(y_1, \ldots, y_n)d\mathbf{y},$$

and the fact that $\mathbf{f}_{\mathbf{X}_i}$ is PF_2 implies $\mathbf{f}_{\mathbf{X}_i}(z_i - y_i)$ is TP_2 in (z_i, y_i).

B. (Markov evolution) If $X_1, \ldots, X_n$ describe the evolution of a Markov chain, which has TP_2 transition probability densities then $\mathbf{X}$ is MTP_2 .

C. (order statistics) If $\mathbf{X}$ is a vector of i.i.d. random variables, which have a density f with respect to σ_i then the vector of order statistics $(X_{(1)}, \ldots, X_{(n)})$ is MTP_2 . The joint density of the order statistics is of the form

$$n!\mathbf{g}(\mathbf{x}) \prod_{i=1}^{n} f(x_i),$$

where

$$\mathbf{g}(\mathbf{x}) = \mathbf{I}_{(x_1 \leq \ldots \leq x_n)}(\mathbf{x}).$$

Since $\mathbf{g}$ is MTP_2 the above density is MTP_2 .

D. (conditional monotone regression) If f is MTP_2 then

$$\mathbf{E}(\phi(X_1,\ldots,X_k) \mid X_{k+1}=x_{k+1},\ldots,X_n=x_n),$$

is a non-decreasing function of $(x_{k+1},\ldots,x_n)$, for each $1 \le k \le n$, whenever ϕ is a nondecreasing function.

E. (states of a Markov process) Let $(\mathbf{X}_t, t \geq 0)$ be a homogeneous Markov process whose state space is $\mathcal{X}$. Let $\mathbf{p}(t,\mathbf{x},\mathbf{y})$ be the transition density of the process with respect to σ . If $\mathbf{X}_0$ is associated and for each fixed t

$$\mathbf{p}(t,\mathbf{x},\mathbf{y}) <_{TP_2} \mathbf{p}(t,\mathbf{x}',\mathbf{y}'),$$

for all $\mathbf{x} \le \mathbf{x}'$, as a function of $\mathbf{y}$ then $\mathbf{X}_t$ is associated.

3.3 A general theory of positive dependence

In this section we recall some general positive dependence notions, which are weaker than association, and which are defined only for R^n (see Shaked (1982)). The general framework of the theory is build up from two equivalent characterizations of association.

- **X** is associated if and only if

  $$\mathbf{P}(\mathbf{X} \in A \cap B) \ge \mathbf{P}(\mathbf{X} \in A)\mathbf{P}(\mathbf{X} \in B),$$

  for all open upper sets A, B (A is upper if $\mathbf{x} \in A$ and $\mathbf{y} \ge \mathbf{x}$ implies $\mathbf{y} \in A$)

- **X** is associated if and only if

  $$Cov(f(\mathbf{X}), g(\mathbf{X})) \ge 0,$$

  for all non-decreasing real functions f, g for which the covariance exists.

Variating the classes from which sets and functions are chosen we introduce several concepts of dependence. Let $\mathcal{A}, \mathcal{B}$ be two collections of sets in R^n .

Definition A. **X** *is positively dependent relative to* $\mathcal{A}$ *and* $\mathcal{B}$ $(PD(\mathcal{A},\mathcal{B}))$ *if*

$$\mathbf{P}(\mathbf{X} \in A \cap B) \ge \mathbf{P}(\mathbf{X} \in A)\mathbf{P}(\mathbf{X} \in B),$$

whenever $A \in \mathcal{A}$ *and* $B \in B$.

Let $\mathcal{F}, \mathcal{G}$ be two families of real n-variate functions.

Definition B. **X** *is functionally positive dependent relative to* $\mathcal{F}$ *and* $\mathcal{G}$ $(FPD(\mathcal{F},\mathcal{G}))$ *if* $Cov(f(\mathbf{X}), g(\mathbf{X})) \ge 0$, *whenever* $f \in \mathcal{F}$ *and* $g \in \mathcal{G}$, *provided the expectations exist.*

Most of the discussion in the literature concentrates around the following collections of sets and functions (see Shaked (1982)).

I. $\mathcal{A}_1$- the collection of open upper orthants in $\mathbf{R}^n$, i.e.

$$A \in \mathcal{A}_1 \Leftrightarrow A = \{\mathbf{x} : x_i > a_i, i = 1, \dots, n\}, a_i \in [-\infty, \infty].$$

II. $\mathcal{A}_2$- the collection of open upper half spaces, i.e.

$$A \in \mathcal{A}_2 \Leftrightarrow A = \{\mathbf{x} : \sum_{i=1}^{n} a_i x_i > c\}, c \in [-\infty, \infty], a_i \geq 0.$$

III. $\mathcal{A}_3$-the collection of sets related to coherent life functions

$$A = \cap_{1 \leq i \leq m} \cap_{j \in C_i} \{\mathbf{x} : x_j > a_j\}$$

or

$$A = \cup_{1 \leq i \leq k} \cap_{j \in D_i} \{\mathbf{x} : x_j > a_j\},$$

where $a_j \in [-\infty, \infty], C_i, D_i \subseteq \{1, \dots, n\}, m, k \in \mathbf{N}_+$.

IV. $\mathcal{A}_4$- the collection of convex open upper sets in R^n.

V. $\mathcal{A}_5$- the collection of open upper sets in R^n .

Functions.

I. $\mathcal{F}_1$- the collection of functions

$$f(\mathbf{x}) = \min_{1 \leq i \leq n} \{b_i x_i\}, b_i \in [0, \infty].$$

II. $\mathcal{F}_2$- the collection of linear functions

$$f(\mathbf{x}) = \sum_{i=1}^{n} a_i x_i, a_i \in [0, \infty].$$

III. $\mathcal{F}_3$- the collection of coherent life functions

$$f(\mathbf{x}) = \min_{1 \leq i \leq m} \max_{j \in C_i} b_j x_j$$

or

$$f(\mathbf{x}) = \max_{1 \leq i \leq k} \min_{j \in D_i} b_j x_j,$$

where $b_j \geq 0, C_i, D_i \subseteq \{1, \dots, n\}, k, m \in \mathbf{N}_+$.

IV. $\mathcal{F}_4$- the collection of concave increasing functions on R^n .

V. $\mathcal{F}_5$- the collection of nondecreasing functions on R^n

If $\mathcal{A} = \mathcal{B}$, we write $PD(\mathcal{A})$, for $PD(\mathcal{A}, \mathcal{A})$. If $\mathcal{F} = \mathcal{G}$ we write $FDP(\mathcal{F})$ for $FDP(\mathcal{F}, \mathcal{F})$

It is not difficult to see that

$$PD(\mathcal{A}_5) \Rightarrow PD(\mathcal{A}_4) \Rightarrow PD(\mathcal{A}_2)$$

$$PD(\mathcal{A}_3) \Rightarrow PD(\mathcal{A}_1)$$

and

$$FDP(\mathcal{F}_5) \Rightarrow FDP(\mathcal{F}_4) \Rightarrow FDP(F_2)$$

$$FDP(F_3) \Rightarrow FDP(F_1).$$

Counterimplications are shown in Shaked (1982).

The notion of $PD(A_j)$ essentially implies the notion of $FDP(\mathcal{F}_j)$.

Lemma C. (Shaked (1982)) *For $j = 1, \ldots, 5$, $\mathbf{X}$ is $PD(\mathcal{A}_j)$ if and only if $(f(\mathbf{X}), g(\mathbf{X}))$ is PQD, whenever $f, g \in F_j$, provided $\mathbf{X}$ is nonnegative.*

A possible generalization of the concept PQD is the following

Definition D.
i) *$\mathbf{X}$ is positively upper orthant dependent (PUOD) if*

$$\mathbf{P}(\mathbf{X} > \mathbf{x}) \geq \prod_{i=1}^{n} \mathbf{P}(X_i > x_i), \mathbf{x} \in \mathbf{R}^n,$$

ii) *$\mathbf{X}$ is positively lower orthant dependent (PLOD) if*

$$\mathbf{P}(\mathbf{X} \leq \mathbf{x}) \geq \prod_{i=1}^{n} \mathbf{P}(X_i \leq x_i), \mathbf{x} \in \mathbf{R}^n.$$

If $\mathbf{X}$ is $PUOD$ and $PLOD$ then we say that it is positively orthant dependent (POD).

It is immediate that $PD(A_1) \Rightarrow PUOD$, so except for $PD(A_2)$ all the set dependence notions are weaker than association, but stronger than the orthant dependence.

Several applications of the above introduced notions, among others distributions with exponential minimums, absolute values of multivariate normal variables, an invariance principle, and bounds on system reliability are studied in detail by Shaked (1982).

3.4 Multivariate orderings and dependence

A nice unified approach to some dependence notions stronger than association is possible via stochastic ordering.

For a vector $\mathbf{X}$ we define sets

$$h_t = \{\mathbf{X}_A = \mathbf{x}_A, \mathbf{X}_{A^c} > t\mathbf{e}\},$$

where $t \geq 0$, $\mathbf{e} = (1, \ldots, 1), A \subseteq \{1, \ldots, n\}$, $0\mathbf{e} \leq \mathbf{X}_A \leq t\mathbf{e}$.
Such sets may represent the history of consecutive failures in some reliability system, where $\mathbf{X}$ denotes the lifetimes of components. Knowing the history up to time t, we are interested what properties possesses the residual life time of components, which are

in the set A^c, i.e., which are still working at time t. Denote the distribution of these residual lives by

$$\mathcal{L}(\mathbf{X}_{A^c} - t\mathbf{e}\,|h_t).$$

For a fixed history h_t, and for each component $i \in A^c$, let (see Lemma 2.2.G.)

$$r_i(h_t) = \lim_{\Delta t \to 0} \frac{1}{\Delta t} P(t < X_i \le t + \Delta t \mid h_t),$$

where $t\mathbf{e} \geq \mathbf{X}_A$. We call $r_i(h_t)$ the conditional hazard rate of X_i at time t, given h_t. We assume that the above limit exists, for all h_t.
The total hazard accumulated by X_i up to time t, given h_t, is defined as follows.
For A with k elements, we denote by $i_1, \ldots, i_k$ the rearrangement of indexes in A, n, for which $x_{i_1} \le \ldots \le x_{i_k}$. Consider the sequence of histories

$$h_t^{i_j} = \{X_{i_1} = x_{i_1}, \ldots, X_{i_j} = x_{i_j}, X_l > t, l \neq i_m, m = 1, \ldots, j\},$$

where $t > x_{i_j}, j = 1, \ldots, k$, and

$$h_t^0 = \{\mathbf{X} > t\mathbf{e}\}.$$

This sequence of histories traces the evolution of consecutive failures in the history h_t.
Let, for $i \in A^c$,

$$R_i(h_t) = \int_0^{x_{i_1}} r_i(h_u^0)du + \int_{x_{i_1}}^{x_{i_2}} r_i(h_u^{i_1})du + \ldots + \int_{x_{i_k}}^{t} r_i(h_u^{i_k})du,$$

where $t \geq x_{i_k}$. $R_i(h_t)$ is the total hazard accumulated by X_i up to time t, given h_t.

Consider two random vectors $\mathbf{X}, \mathbf{Y}$, with the corresponding conditional hazard rates $r_i(h_t)$, $q_i(h_t)$, and the total hazard rates, given h_t, $R_i(h_t)$, $Q_i(h_t)$, respectively ($i = 1, \ldots, n$). We write $h_t > \hat{h}_t$ if the failures in h_t are more numerous, i.e. $\hat{A} \subseteq A$, and for the components which failed in both histories, the failures in h_t are earlier than the failures in $\hat{h}_t$, that is, for $i \in \hat{A}$, $x_i \le \hat{x}_i$.

Definition A. *For two random vectors* $\mathbf{X}, \mathbf{Y}$, *we write* $\mathbf{X} <_h \mathbf{Y}$ *if*

$$r_i(h_t) \geq q_i(\hat{h}_t),$$

for all $h_t > \hat{h}_t$, $i \in A^c$.

Definition B. *For two random vectors* $\mathbf{X}, \mathbf{Y}$, *we write* $\mathbf{X} <_{ch} \mathbf{Y}$ *if*

$$R_i(h_t) \geq Q_i(\hat{h}_t),$$

for all $h_t > \hat{h}_t, i \in A^c$.

The above introduced relations are not reflexive (they are pre-orders). Of course $\mathbf{X} <_h \mathbf{Y}$ implies $\mathbf{X} <_{ch} \mathbf{Y}$.

Theorem C. (Shaked and Shanthikumar (1987)(1989)) *The following implications hold true*

$$\mathbf{X} <_{TP} \mathbf{Y} \Rightarrow \mathbf{X} <_{h} \mathbf{Y} \Rightarrow \mathbf{X} <_{ch} \mathbf{Y} \Rightarrow \mathbf{X} <_{st} \mathbf{Y}.$$

From the first section, we know that $\mathbf{X} <_{TP_2} \mathbf{X}$ is equivalent to $\mathbf{X}$ be MTP_2. In this way, using $<_h$, $<_{ch}$, we can similarly define further notions of positive dependence, which are weaker than MTP_2, but stronger than association.

Definition D. -
a) *We say that* $\mathbf{X}$ *has hazard rates increasing by failures (HIF) if*

$$\mathbf{X} <_h \mathbf{X};$$

b) *The vector* $\mathbf{X}$ *have supportive lifetimes (SL) if*

$$\mathbf{X} <_{ch} \mathbf{X}.$$

Another notion of positive dependence, which is related to the above ones is WBF (weakened by failures).

Definition E. *The vector* $\mathbf{X}$ *is weakened by failures (WBF) if*

$$\mathcal{L}(\mathbf{X}_{A^c} - t\mathbf{e}\ |h_t^{+i}) <_{st} \mathcal{L}(\mathbf{X}_{A^c} - t\mathbf{e}\ |h_t),$$

where

$$h_t^{+i} = \{\mathbf{X}_A = \mathbf{x}_A, X_i = t, \mathbf{X}_{A^c - \{i\}} > t\mathbf{e}\},\ \ i \in A^c,\ \ t \geq \max\{x_i, i \in A\}.$$

Norros (1986) showed that $SL \Rightarrow WBF$, and Arjas and Norros (1984) showed that $WBF \Rightarrow$ association.

The notions HIF and SL can also be characterized by distributions of residual life times.

Theorem F. (Shaked and Shanthikumar (1989)) *a)* $\mathbf{X}$ *is* SL *if and only if*

$$\mathcal{L}(\mathbf{X}_{A^c} - t\mathbf{e}\ |h_t^{+i}) <_{ch} \mathcal{L}(\mathbf{X}_{A^c} - t\mathbf{e}\ |h_t),$$

b) $\mathbf{X}$ *is HIF if and only if*

$$\mathcal{L}(\mathbf{X}_{A^c} - t\mathbf{e}\ |h_t^{+i}) <_h \mathcal{L}(\mathbf{X}_{A^c} - t\mathbf{e}\ |h_t),$$

where $i \in A^c,\ t > \max\{x_i, i \in A\}$.

3.5 Negative association

In many cases, for distributions with nonpositive correlations, it is important to have checkable conditions, which imply the inequality

$$P(X_1 > x_1, \ldots, X_n > x_n) \leq \prod_{i=1}^{n} P(X_i > x_i),$$

for $x_i \in R^n$. An $\mathbf{X}$ satisfying such a condition is called negatively upper orthant dependent ($NUOD$) (with the reversed inequalities inside brackets we define $NLOD$). Some concepts of negative dependence, which are in a sense reverse to the MTP_2 notion were developed by Karlin and Rinott (1980), and modified in Block et al. (1982).

Let μ be a probability measure on the Borel sets in R^n. For intervals $I_1, \ldots, I_k$ in R define a set function

$$\mu(I_1, \ldots, I_n) = \mu(I_1 \times \ldots \times I_n).$$

For intervals I, J in R write $I < J$ if $x \in I, y \in J$ implies x <y.

Definition A. *For a probability measure on R^n we say that μ is RR_2 in pairs if $\mu(I_1, \ldots, I_n)$ is RR_2 in all pairs of variables, when the remaining variables are held fixed.*

Note that we do not need assume the existence of a density in the above definitions.

A vector $\mathbf{X}$ is said to be RR_2 in pairs if its distribution is RR_2 in pairs.

Lemma B. (Block et al. (1982)) *If $\mathbf{X}$ is RR_2 in pairs then*

$$P(\mathbf{X} > \mathbf{x}) \leq \prod_{i=1}^{n} P(X_i > x_i),$$

and

$$P(\mathbf{X} \leq \mathbf{x}) \leq \prod_{i=1}^{n} P(X_i \leq x_i),$$

for $x_i \in R$ (i.e. $\mathbf{X}$ is negatively orthant dependent (NOD)).

The following results from Block et al. (1982) are useful for constructing new negatively dependent distributions from the known ones.

Theorem C.

(i) *If $\mathbf{X}$ is RR_2 in pairs then $(\phi_1(X_1), \ldots, \phi_n(X_n))$ is RR_2 in pairs;*

(ii) *If $\mathbf{X}, \mathbf{Y}$ are independent, and both RR_2 in pairs then $(\mathbf{X}, Y)$ is RR_2 in pairs.*

Before we introduce another stronger notion of negative dependence recall that an univariate density function f is said to be a Polya frequency function of order 2 (PF_2) if $f(x-y)$ is TP_2 on $R \times R$. A probability function f is PF_2 if $f(x-y)$ is TP_2 on $\mathbf{Z} \times \mathbf{Z}$, where $\mathbf{Z}$ denotes the set of integers.

Definition D. $\mathbf{X}$ *satisfies condition N if there exist* $n+1$ *independent random variables* $Y_0, Y_1, \ldots, Y_n$*, each with* PF_2 *density, and a real number s such that*

$$\mathbf{X} =^{st} [(Y_1, \ldots, Y_n) \mid Y_0 + \ldots + Y_n = s]$$

Theorem E. (Block et al. (1982)) *If* $\mathbf{X}$ *satisfies condition N then* $\mathbf{X}$ *is* RR_2 *in pairs.*

Standard examples of distributions, which are considered to be negative dependent in some sense, often satisfy the structural condition N.

Examples.

1. (multinomial) Let $\mathbf{X}$ have the joint probability function with parameters $(N, p_1, \ldots, p_n)$

$$P(\mathbf{X} = \mathbf{x}) = [N!/x_1! \ldots x_n!(N - \sum_{i=1}^{n} x_i)!] \prod_{i=1}^{n} p_i^{x_i} \times (1 - \sum_{i=1}^{n} p_i)^{N - \sum_{i=1}^{n} x_i},$$

where $x_i \geq 0, \sum_{i=1}^n x_i \leq N, p_i \geq 0, 0 < \sum_{i=1}^n p_i < 1$.
The multinomial distribution is the conditional distribution of independent Poisson random variables given their sum. From the last theorem $\mathbf{X}$ is RR_2 in pairs.

2. (multivariate hypergeometric) Let $\mathbf{X}$ have the probability function

$$P(\mathbf{X} = \mathbf{x}) = \binom{M}{N}^{-1} [\prod_{i=1}^{n} \binom{M_i}{x_i}] \binom{M - \sum_{i=1}^n M_i}{N - \sum_{i=1}^n x_i},$$

$x_i \geq 0, \sum_{i=1}^n x_i \leq N, \sum_{i=1}^n M_i \leq M, N, M, M_i$ are positive integer valued parameters. The multivariate hypergeometric distribution is the conditional distribution of independent binomial random variables given their sum. Hence $\mathbf{X}$ is RR_2 in pairs.

Further examples of RR_2 in pairs distributions are the Dirichlet, Dirichlet compound multinomial (see Block at al (1982)).

Another concept of negative dependence is the notion of negatively associated (NA) random variables. The theory and application of NA are not simply the duals of the theory and application of positive association, but differ in important respects. Negative association has one distinct advantage over the other known types of negative dependence. Increasing functions of disjoint sets of NA random variables are also NA. This type of closure property does not hold, for example for RR_2 in pairs.

We formulate first a technical lemma for the conditional covariance, defined by

$$Cov(X, Y \mid \mathcal{F}) = E(XY \mid \mathcal{F}) - E(X \mid \mathcal{F})E(Y \mid \mathcal{F}),$$

where $\mathcal{F}$ is an arbitrary σ-field.

Lemma F. *Let* (X, Y) *be a pair of real random variables on* Ω*, and* $\mathcal{F}_1 \subseteq \mathcal{F}_2$ *two* σ*-fields on* Ω*. Then*

$$Cov(X, Y \mid \mathcal{F}_1) = E(Cov(X, Y \mid \mathcal{F}_2) \mid \mathcal{F}_1) + Cov(E(X \mid F_2), E(Y \mid \mathcal{F}_2) \mid \mathcal{F}_1).$$

In particular, when $\mathcal{F}_1 = \{\Omega, \emptyset\}$

$$Cov(X, Y) = E(Cov(X, Y \mid \mathcal{F}_2)) + Cov(E(X \mid F_2), E(Y \mid \mathcal{F}_2)).$$

Proof. By rewriting of the right-hand side of the equality we obtain

$$E(E(XY \mid \mathcal{F}_2) - E(X \mid \mathcal{F}_2)E(Y \mid \mathcal{F}_2) \mid \mathcal{F}_1) + E(E(X \mid \mathcal{F}_2)E(Y \mid \mathcal{F}_2) \mid \mathcal{F}_1) -$$

$$-E(E(X \mid \mathcal{F}_2) \mid \mathcal{F}_1)E(E(Y \mid \mathcal{F}_2) \mid \mathcal{F}_1) = E(XY \mid \mathcal{F}_1) - E(X \mid \mathcal{F}_1)E(Y \mid \mathcal{F}_1) =$$
$$= Cov(X, Y \mid \mathcal{F}_1),$$

since from the assumption that $\mathcal{F}_1 \subseteq \mathcal{F}_2$,

$$E(E(XY \mid F_2) \mid \mathcal{F}_1) = E(XY \mid \mathcal{F}_1).$$

□

Definition G. **X** *is negatively associated (NA) if, for every subset* $A \subseteq \{1, \dots, n\}$

$$Cov(f(X_i, i \in A), g(X_j, j \in A^c)) \le 0,$$

whenever f, g *are nondecreasing.*

NA may also refer to the set of random variables $\{X_1, \dots, X_n\}$, or to the underlying distribution of **X**.

Negative association possesses the following properties (see Joag-Dev and Proschan (1983))

P1. A pair (X, Y) of random variables is NA if and only if

$$P(X \le x, Y \le y) \le P(X \le x)P(Y \le y),$$

i.e. (X, Y) is negatively quadrant dependent (NQD).

P2. For disjoint subsets $A_1, \dots, A_m$ of $\{1, \dots, n\}$, and nondecreasing positive functions $f_1, \dots, f_m$, **X** is NA implies

$$E \prod_{i=1}^{m} f_i(\mathbf{X}_{A_i}) \le \prod_{i=1}^{m} E f_i(\mathbf{X}_{A_i}),$$

where $\mathbf{X}_{A_i} = (X_j, j \in A_i)$.

P3. If **X** is NA then it is NOD.

P4. A subset of NA random variables is NA.

P5. If **X** has independent components then it is NA.

P6. Increasing functions defined on disjoint subsets of a set of NA random variables are NA.

P7. If **X** is NA and **Y** is NA, and **X** is independent of **Y** then $(\mathbf{X}, \mathbf{Y})$ is NA.

In some applications negative association is created, when the random variables are subjected to conditioning.

Theorem H. *Let $X_1, \ldots, X_n$ be independent, and suppose that*

$$E(f(\mathbf{X}_A) \mid \sum_{i \in A} X_i) \tag{1}$$

is increasing in $\sum_{i \in A} X_i$, for every nondecreasing f, and every $A \subseteq \{1, \ldots, n\}$. Then the conditional distribution of $(\mathbf{X} \mid \sum_{i=1}^{n} X_i = s)$ is NA, for almost all s.

Proof. Let $A \subseteq \{1, \ldots, n\}$

$$S_1 = \sum_{i \in A} X_i, \qquad S_2 = \sum_{i \in A^c} X_i, \qquad S = S_1 + S_2,$$

and f, g be nondecreasing real functions. Denote by $\mathbf{X}^*$ a vector with the distribution, which equals to the conditional distribution of $(\mathbf{X} \mid S)$. It is clear from the definition of conditional covariance that, for arbitrary measurable functions ϕ, ψ,

$$Cov(\phi(\mathbf{X}^*), \psi(\mathbf{X}^*)) = Cov(\phi(\mathbf{X}), \psi(\mathbf{X}^*) \mid S),$$

so

$$Cov(f(\mathbf{X}^*_A), g(\mathbf{X}^*_{A^c})) = Cov(f(\mathbf{X}_A), g(\mathbf{X}_{A^c}) \mid S).$$

Using Lemma F., where conditioning is taken for $\mathcal{F}_1 = \sigma(S)$, $\mathcal{F}_2 = \sigma(S_1, S_2)$, we have from the independence between $\mathbf{X}_A$ and $\mathbf{X}_{A^c}$

$$Cov(f(\mathbf{X}_A), g(\mathbf{X}_{A^c}) \mid S) = Cov(E(f(\mathbf{X}_A) \mid S_1, S_2), E(g(\mathbf{X}_{A^c}) \mid S_1, S_2) \mid S).$$

Now, for $S = s$ this can be rewritten as

$$Cov(\phi_A(S_1), \psi_{A^c}(S_1)),$$

where from (1), ϕ_A is nondecreasing, and ψ_{A^c} is nonincreasing. Since the above covariance is certainly nonpositive (single random variable is associated), we have, for almost all s

$$Cov(f(\mathbf{X}^*_A), g(\mathbf{X}^*_{A^c})) \leq 0,$$

i.e. the distribution of $(\mathbf{X} \mid S = s)$ is NA. □

The above theorem takes on added interest, when considered in conjunction with the following theorem

Theorem I. (Efron (1965)) *Let $X_1, \ldots, X_n$ be mutually independent with PF_2 densities. Then*

$$E(\phi(\mathbf{X}) \mid S = s)$$

is increasing in (almost every) s, provided ϕ is nondecreasing.

Corollary J. *If $X_1, \ldots, X_n$ are independent with PF_2 densities then the conditional distribution of $(\mathbf{X} \mid S = s)$ is NA, for almost all s.*

NA also arises naturally via permutation distributions.

PROBLEMS AND REMARKS

A. Let $\mathbf{x} = (x_1, \ldots, x_n)$ be a set of real numbers. A permutation distribution is the joint distribution of the vector $\mathbf{X}$, which takes as values all permutations of $\mathbf{x}$ with equal probabilities $1/n!$. Such a distribution is NA (see Joag-Dev and Proschan (1983)).

B. (multinomial) We already know that the multinomial distribution is RR_2 in pairs. That it is also NA can be seen from the above corollary since it is the conditional distribution of independent Poisson random variables given their sum. Another way to derive NA for this distribution is possible via the property $P6$. Let $\mathbf{Z} = (Z_1, \ldots, Z_n)$ be a vector having a multinomial distribution, obtained by taking only one observation. Thus only one Z_i is 1 while the rest are zero. The NA property of $\mathbf{Z}$ trivially follows from the definition. A general multinomial distribution is a convolution of independent copies of $\mathbf{Z}$, the property $P6$ establishes NA in this case.

C. (multivariate hypergeometric) We know that the multivariate hypergeometric distribution is RR_2 in pairs. From the above corollary it is also NA. We can see this from the property $P6$. An urn contains M balls each having a different color. Suppose a random sample of N balls is chosen (without replacement) and Y_i, indicate the presence of ball of the ith color in the sample. Clearly $\mathbf{Y}$ has a permutation distribution, and hence is NA. More generally, M_i balls are of the ith color, $i = 1, \ldots, n$, with $\sum_{i=1}^{n} M_i = M$, and let X_i be the number of balls of the ith color in the sample. Then X_i can be viewed as the sum of M_i indicators in the simple model above. Since the X_i are sums over nonoverlapping sets of random variables, by $P6$, NA is transmitted.

3.6 Independence via uncorrelatedness

The classes of multivariate distributions defined by all of the notions of positive or negative dependence, for $\mathbf{X}$, invariably contain those where X_i's are mutually independent. It is of some interest to see whether, in such classes, the simple condition of uncorrelatedness characterizes mutual independence.

Let A, B be subsets of $\{1, \ldots, n\}$. Define

$$\mathbf{H}_{\mathbf{X}_A, \mathbf{X}_B}(\mathbf{x}) = P(X_j > x_j, j \in A \cup B) - P(X_j > x_j, j \in A) P(X_j > x_j, j \in B).$$

We start with a technical lemma

Lemma A. (Lebowitz (1972)) *If* $\mathbf{X}$ *is associated then, for arbitrary* $A, B \subseteq \{1, \ldots, n\}$, *and* $\mathbf{x} \in \mathbf{R}^n$

$$0 \le \mathbf{H}_{\mathbf{X}_A, \mathbf{X}_B}(\mathbf{x}) \le \sum_{i \in A} \sum_{j \in B} \mathbf{H}_{\mathbf{X}_i, \mathbf{X}_j}(x_i, x_j).$$

If $\mathbf{X}$ is NA then, for disjoint A, B the above inequalities are reversed.

Proof. Define, for a fixed $\mathbf{x}$

$$U(A) = \prod_{i \in A} I_{(X_i > x_i)},$$

$$V(A) = \sum_{i \in A} I_{(X_i > x_i)}.$$

Note that $U(A), V(A)$ and $V(A) - U(A)$ are increasing functions of $\mathbf{X}$, so

$$Cov(U(A), V(B) - U(B)) \geq 0, Cov(V(A) - U(A), V(B)) \geq 0,$$

and we have

$$0 \leq Cov(U(A), U(B)) \leq Cov(U(A), U(B)) + Cov(U(A), V(B) - U(B)) =$$
$$= Cov(U(A), V(B)) \leq Cov(U(A), V(B)) + Cov(V(A) - U(A), V(B)) =$$
$$= Cov(V(A), V(B)).$$

This is the required inequality since

$$\mathbf{H}_{\mathbf{X}_A, \mathbf{X}_B}(\mathbf{x}) = Cov(U(A), U(B)),$$

and

$$\sum_{i \in A} \sum_{j \in B} \mathbf{H}_{\mathbf{X}_i, \mathbf{X}_j}(x_i, x_j) = Cov(V(A), V(B)).$$

In the NA case the inequalities can be reversed, for disjoint A, B. □

Corollary B. *Suppose* $\mathbf{X}$ *is associated or negatively associated. Then*
a) $\mathbf{X}_A$ *is independent of* $\mathbf{X}_B$ *if and only if* $Cov(X_i, X_j) = 0$*, for* $i \in A, j \in B$, $A \cap B = \emptyset$.
b) $X_1, \ldots, X_n$ *are mutually independent if and only if* $Cov(X_i, X_j) = 0$*, for all* $i \neq j$.

Proof. We know from the first section of this chapter (see Theorem 3.1.E.), that the assumption $\mathrm{Cov}(X_i, X_j) = 0$ implies that X_i, X_j are pairwise independent. From the above theorem it follows $\mathbf{H}_{\mathbf{X}_A, \mathbf{X}_B}(\mathbf{x}) = 0$, for every pair of disjoint subsets A, B, and arbitrary $\mathbf{x}$. Thus we have that $\mathbf{X}_A, \mathbf{X}_B$ are independent, and $X_1, \ldots, X_n$ are mutually independent. □

The second part of the last corollary can also be shown to be valid under weaker hypotheses on the dependence of $\mathbf{X}$ than association or NA. We define further notions of dependence.

Definition C. (Joag-Dev (1983)) $\mathbf{X}$ *is said to be strongly positive orthant dependent (SPOD) if, for every* $A \subseteq \{1, \ldots, n\}$*, and* $\mathbf{x}$

$$P(\mathbf{X} > \mathbf{x}) \geq P(\mathbf{X}_A > \mathbf{x}_A) P(\mathbf{X}_{A^c} > \mathbf{x}_{A^c}),$$

$$P(\mathbf{X} \leq \mathbf{x}) \geq P(\mathbf{X}_A \leq \mathbf{x}_A) P(\mathbf{X}_{A^c} \leq \mathbf{x}_{A^c}),$$

and

$$P(\mathbf{X}_A > \mathbf{x}_A, \mathbf{X}_{A^c} \leq \mathbf{x}_{A^c}) \leq P(\mathbf{X}_A > \mathbf{x}_A) P(\mathbf{X}_{A^c} \leq \mathbf{x}_{A^c}).$$

Analogously we define the corresponding negative dependence ($SNOD$).

Definition D. **(Joag-Dev (1983))** **X** *is said to be linearly positively quadrant dependent (LPQD) if, for every pair of nonnegative vectors* $\mathbf{r}, \mathbf{s}$, *and for every* $A \subseteq \{1, \ldots, n\}$, *the pair*

$$(\sum_{i \in A} r_i x_i, \sum_{j \in A^c} s_j x_j)$$

is PQD.

Note that $LPQD$ is weaker than $PD(\mathcal{A}_2)$ introduced by Shaked (1982). Also neither of the two conditions $SPOD$ and $LPQD$ implies the other.

We have however

Theorem E. *Suppose that* **X** *fulfills one of the conditions SPOD, SNOD, LPQD, LNQD. It follows that* **X** *has mutually independent components if and only if* $Cov(X_i, X_j) = 0$, *for all* $i \neq j$.

Proof. We consider first the case $SPOD$ ($SNOD$). The proof in this case is by induction. For $n = 3$, let

$$p_{110} = P(I_{(X_1 > x_1)} = 1, I_{(X_2 > x_2)} = 1, I_{(X_3 > x_3)} = 0)$$

etc. Note that from the uncorrelatedness and SPOD, the vector (X_1, X_2, X_3) is pairwise independent. Now, for all triplets, which contain both 0 and 1, using pairwise independence, and the definition of $SPOD$, we have inequalities of the type

$$p_{101} \leq p_1(1 - p_2)p_3,$$

$$p_{001} \leq (1 - p_1)(1 - p_2)p_3, etc.,$$

where $p_i = P(X_i > x_i)$. However, these have to be equalities, because if not, combining the two above it would follow

$$P(I_{(X_2 > x_2)} = 0, I_{(X_3 > x_3)} = 1) < (1 - p_2)p_3,$$

violating the pairwise independence.
The reverse inequalities we have for

$$p_{111} \geq p_1 p_2 p_3,$$

$$p_{000} \geq (1 - p_1)(1 - p_2)(1 - p_3).$$

But again, these have to be equalities since the sum of the right and left sides of all these expressions has to be 1.
For the induction step, assume that every subset of cardinality $n - 1$ consists of mutually independent random variables. This leads to inequalities which are similar to the above ones. We then proceed analogously.
Consider now the case $LPQD$ ($LNQD$). Similarly to (3), we obtain

$$Cov(e^{irX}, e^{isY}) = \int_{-\infty}^{\infty} \int_{-\infty}^{\infty} ire^{irx} ise^{isy} H_{X,Y}(x, y) dx dy.$$

For (X, Y), which is PQD (NQD), we then have

$$\mid E(e^{irX+isY}) - E(e^{irX})E(e^{isY}) \mid \leq \mid rsCov(X,Y) \mid,$$

for all real r,s. In terms of the characteristic functions $\Phi_X(r) = E(e^{irX})$, and $\Phi_{X,Y}(r,s) = E(e^{irX+isY})$, it means

$$\mid \Phi_{X,Y}(r,s) - \Phi_X(r)\Phi_Y(s) \mid \leq \mid rsCov(X,Y) \mid .$$

This can be generalized by induction (see Newman (1984)) to

$$\mid \Phi_{\mathbf{X}}(r_1, \ldots, r_n) - \prod_{j=1}^{n} \Phi_{X_j}(r_j) \mid \leq \sum_{i<j} \mid r_i r_j Cov(X_i, X_j) \mid,$$

provided $\mathbf{X}$ is $LPQD$ ($LNQD$).
The case $LPQD$ ($LNQD$) of our theorem is an immediate consequence of the above inequality. □

3.7 Association for Markov processes

Consider the state space $E = \mathcal{X}_1 \times \ldots \times \mathcal{X}_n$, where each $\mathcal{X}_i$, is a Polish space, and E is a normally ordered Polish space with a closed partial order $\prec$. The term "normally ordered" we understand in the sense of Lindqvist (1988) (see Section 2.4), i.e. for every pair of disjoint sets, $F_1 = \{x : x \prec y \text{ for some } y \in K_1\}$, $F_2 = \{x : y \prec x \text{ for some } y \in K_2\}$, where K_1, K_2 are compact subsets of E, there exists an increasing, continuous function f with $f(x) = 0$ for all $x \in F_1, f(x) = 1$ for all $x \in F_2$, and $0 \leq f(x) \leq 1$ for all $x \in E$. For instance, $E = \mathbf{R}^n$, or $E = \mathbf{Z}^n$ with the coordinatewise order, which we always denote by $\leq$. This concept is slightly weaker then the one defining a normally ordered space by Nachbin(1965), however it proved to be useful for non compact spaces (see Lindqvist (1988)). We need this notion in order to characterize correlated probability measures by continuous and bounded functions.

In the case we need a linear structure on E we always assume that the order and the topology are compatible with the addition and multiplication.

By a partition $\mathcal{P}$ of the set $\{1, \ldots, n\}$ we understand a class of disjoint subsets (segments) of this set which in common contain all of its elements. For each segment J of a partition $\mathcal{P}$ we consider a closed partial order $\prec_J$ on $\mathbf{X}_{i\in J}\mathcal{X}_i$, $J \in \mathcal{P}$, such that the space $(\mathbf{X}_{i\in J}\mathcal{X}_i, \prec_J)$ is normally ordered. These orders need not be of the same kind on each segment, for example we may have the coordinatewise order on one segment while on some other the partial sum order, etc. We will refer to such orders as to **partition separated orders**. Such a collection of orders defines for a given partition $\mathcal{P}$ an order $\prec_{(\mathcal{P})}$ on the product space E by

$$\mathbf{x} \prec_{(\mathcal{P})} \mathbf{y} \text{ iff } \mathbf{x}_J \prec_J \mathbf{y}_J, J \in \mathcal{P},$$

where $\mathbf{x}_J$ is the projection of $\mathbf{x}$ onto $\mathbf{X}_{i\in J}\mathcal{X}_i$, $J \in \mathcal{P}$.

The set of all continuous, bounded real functions increasing with respect to $\prec_{(\mathcal{P})}$ we denote by $I^*_{(\mathcal{P})}$. If each $\prec_J$ is the coordinatewise order then $\prec_{(\mathcal{P})}$ reduces to the coordinatewise order $\leq$ on E.

Denote by $\mathcal{M}$ the set of probability measures on E equipped with the Borel σ -algebra. Together with a given partition $\mathcal{P}$ we consider the following stochastic ordering for measures $\mathbf{P}, \mathbf{Q} \in \mathcal{M}$.

$$\mathbf{P} \prec_{st\mathcal{P}} \mathbf{Q} \text{ if } \int_E f d\mathbf{P} \leq \int_E f d\mathbf{Q}$$

for all $f \in I^*_{(\mathcal{P})}$, for which the integrals exist.

The $\prec_{st\mathcal{P}}$ is called the strong stochastic ordering generated by $\prec_{(\mathcal{P})}$.

We say that $\mathbf{P} \in \mathcal{M}$ is associated-($\prec_{(\mathcal{P})}$) if

$$\int_E fg d\mathbf{P} \geq \int_E f d\mathbf{P} \int_E g d\mathbf{P}$$

for all $f, g \in I^*_{(\mathcal{P})}$ for which the integrals exist.

Consider a Markov process $\mathbf{X} = (\mathbf{X}(t), t \geq 0)$ with values in E. Assume that it is a Feller process, i.e. $\mathbf{T}(t)f \in C(E)$ if $f \in C(E)$, where $C(E)$ denotes the class of continuous, bounded real functions, and $\mathbf{T} = (\mathbf{T}(t), t \geq 0)$ is the corresponding Markov semigroup. Assume also that $\mathbf{T}$ is strongly continuous. Denote by $\mathbf{A}$ the generator of $\mathbf{T}$, with domain $D_{\mathbf{A}}$. Note that $D_{\mathbf{A}} \cap I^*_{(\mathcal{P})}$ is dense in $I^*_{(\mathcal{P})}$. Indeed,

$$f \in I^*_{(\mathcal{P})} \Rightarrow \int_0^h \mathbf{T}_s f ds \in D_{\mathbf{A}} \cap I^*_{(\mathcal{P})},$$

and

$$\lim_{h \to 0} \frac{1}{h} \int_0^h \mathbf{T}_s f ds = f.$$

For a given $\mu = \mathbf{P}^{\mathbf{X}_0}$ we denote by $\mu\mathbf{T}(t)$ the distribution of $\mathbf{X}$ at time t with the initial distribution μ.

With a partition $\mathcal{P}$ defined for E we introduce a partition $\mathcal{P}^k$ for E^k, which divides the set $\{1, \ldots, nk\}$ segmentwise: each segment $\{ni + 1, \ldots, (i + 1)n\}$ $(i = 0, \ldots, k - 1)$, is partitioned by $\mathcal{P}$.

For a given partition $\mathcal{P}$ and an ordering $\prec_{(\mathcal{P})}$ we say that $\mathbf{X}$ is associated-($\prec_{(\mathcal{P})}$) in time if each collection $(\mathbf{X}(t_1), \ldots, \mathbf{X}(t_k))$ as a random vector with values in E^k is associated-($\prec_{(\mathcal{P}^k)}$), $t_1 < \ldots < t_k$, $k > 1$.

Theorem A. (Liggett(1985)) *Suppose that for each $f \in I^*_{(\mathcal{P})}$ we have $\mathbf{T}(t)f \in I^*_{(\mathcal{P})}$, for all $t \geq 0$. If*

$$\mathbf{A}fg \geq f\mathbf{A}g + g\mathbf{A}f$$

for all $f, g \in D_{\mathbf{A}} \cap I^*_{(\mathcal{P})}$, then $\mathbf{X}$ is associated-($\prec_{(\mathcal{P})}$) in time provided $\mathbf{X}(0)$ has a distribution μ which is associated-($\prec_{(\mathcal{P})}$).

We stated here only the "if" part of the theorem, which we need in this paper; we sketch the main steps of the proof for this part. The formulation in Liggett(1985) was for $\mathcal{P} = \{1, \ldots, n\}$, and bounded $\mathbf{A}$, but it does not change the argument essentially.
Proof. Let $f, g \in D_{\mathbf{A}} \cap I^*_{(\mathcal{P})}$, and

$$F(t) = \mathbf{T}(t)fg - [\mathbf{T}(t)f][\mathbf{T}(t)g].$$

We have

$$F'(t) = \mathbf{A}\mathbf{T}(t)fg - [\mathbf{A}\mathbf{T}(t)f][\mathbf{T}(t)g] - [\mathbf{T}(t)f][\mathbf{A}\mathbf{T}(t)g].$$

From the assumptions of the theorem applied to $\mathbf{T}(t)f$, $\mathbf{T}(t)g$ which are in $D_{\mathbf{A}} \cap I^*_{(\mathcal{P})}$ again

$$F'(t) \geq \mathbf{A}\mathbf{T}(t)fg - \mathbf{A}\{[\mathbf{T}(t)f][\mathbf{T}(t)g]\} = \mathbf{A}F(t)$$

hence from the assumption on $\mathbf{A}$

$$F'(t) = \mathbf{A}F(t) + G(t), \quad F(0) = 0,$$

for some $G(t) \geq 0$. The solution of this problem is

$$F(t) = \mathbf{T}(t)F(0) + \int_0^t \mathbf{T}(t-s)G(s)ds, \quad t \geq 0$$

which is nonnegative since $G(t) \geq 0$. (For a proof see Theorem 2.15 of Chapter I, Liggett (1985), which applies in our setting with obvious modifications in the proof). This implies

$$\mathbf{T}(t)fg \geq [\mathbf{T}(t)f][\mathbf{T}(t)g], \qquad f, g \in D_{\mathbf{A}} \cap I^*_{(\mathcal{P})}. \tag{1}$$

Because $D_{\mathbf{A}} \cap I^*_{(\mathcal{P})}$ is dense in $I^*_{(\mathcal{P})}$ and $\mathbf{T}(t)$ is bounded this inequality extends to $I^*_{(\mathcal{P})}$. The initial distribution μ is associated($\prec_{(\mathcal{P})}$) , so

$$\int_E [\mathbf{T}(t)f][\mathbf{T}(t)g]d\mu \geq \int_E \mathbf{T}(t)fd\mu \int_E \mathbf{T}(t)gd\mu.$$

Combining this with (1) gives association($\prec_{(\mathcal{P})}$) of $\mu\mathbf{T}(t)$.
The rest of the proof is by induction, analogously to the proof of Corollary 2.24 in Chapter II of Liggett(1985). □

Note that Liggett's condition $\mathbf{A}fg \geq f\mathbf{A}g + g\mathbf{A}f$ is equivalent to $\mathbf{T}(t)fg \geq [\mathbf{T}(t)f][\mathbf{T}(t)g]$ for $f, g \in I^*_{(\mathcal{P})}$ which means that each measure $\mathbf{P}_t(\mathbf{x}, .)$ is associated($\prec_{(\mathcal{P})}$) for the corresponding transition kernels (see Lindqvist (1988) for associated kernels).

Unfortunately for many complex Markovian models, the Liggett condition is not fulfilled. However some positive dependence properties can be established by the following results.

Theorem B. (Daduna and Szekli (1992)) *Suppose that* $\mathbf{T}(t)f \in I^*_{(\mathcal{P})}$, *for each* $f \in I^*_{(\mathcal{P})}$, *and* $t \geq 0$. *If* μ *is a stationary measure for* $\mathbf{X}$ *and is associated-*($\prec_{(\mathcal{P})}$), *then*

$$\mathbf{E}_\mu(f_1(\mathbf{X}(t_1)) \ldots f_k(\mathbf{X}(t_k)) \geq \mathbf{E}_\mu f_1(\mathbf{X}(t_1)) \ldots \mathbf{E}_\mu f_k(\mathbf{X}(t_k)),$$

for all nondecreasing $f_i, i = 1, \ldots, k$, *and* $t_1 < \ldots < t_k, k \geq 1$.
Hence, every two random variables $(f_i(\mathbf{X}(t_i)), f_j(\mathbf{X}(t_j))), i, j \in \{1, \ldots, k\}$, *are PQD when the process is in equilibrium .*

Proof. Because μ is a stationary measure for $\mathbf{X}$ we have

$$\mathbf{E}_\mu f(\mathbf{X}(0)) = \mathbf{E}_\mu f(\mathbf{X}(t)), \tag{2}$$

for all $t \geq 0$, and similarly

$$\mathbf{E}_\mu(f(\mathbf{X}(0))g(\mathbf{X}(s))) = \mathbf{E}_\mu(f(\mathbf{X}(t))g(\mathbf{X}(t+s))), \tag{3}$$

for all measurable $f, g, t, s \geq 0$.
We proceed by induction on k. For $k = 2$

$$\mathbf{E}_\mu(f(\mathbf{X}(0))g(\mathbf{X}(s))) = \int \mu(d\mathbf{x})f(\mathbf{x}) \int \mathbf{P}_s(\mathbf{x}, d\mathbf{y})g(\mathbf{y}) = \int \mu(d\mathbf{x})f(\mathbf{x})\tilde{g}(\mathbf{x}),$$

where $\tilde{g} = \mathbf{T}(s)g \in I^*_{(\mathcal{P})}$ from the assumption.
The association of μ yields

$$\int \mu(d\mathbf{x})f(\mathbf{x})\tilde{g}(\mathbf{x}) \geq \int \mu(d\mathbf{x})f(\mathbf{x}) \int \mu(d\mathbf{x})\tilde{g}(\mathbf{x}).$$

Hence

$$\mathbf{E}_\mu(f(\mathbf{X}(0))g(\mathbf{X}(s))) \geq \mathbf{E}_\mu f(\mathbf{X}(0))\mathbf{E}_\mu g(\mathbf{X}(s)).$$

Similarly, from (3) we have

$$\mathbf{E}_\mu(f(\mathbf{X}(t_1))g(\mathbf{X}(t_2))) \geq \mathbf{E}_\mu f(\mathbf{X}(t_1))\mathbf{E}_\mu g(\mathbf{X}(t_2)),$$

for arbitrary $0 \leq t_1 < t_2$, and nondecreasing f, g. Suppose the inequality is valid for $k-1$. We have

$$\mathbf{E}_\mu(f_1(\mathbf{X}(t_1)) \dots f_k(\mathbf{X}(t_k))) =$$
$$= \int \mu(d\mathbf{x}) \int \mathbf{P}_{t_1}(\mathbf{x}, d\mathbf{x}_1)f_1(\mathbf{x}_1) \dots \int \mathbf{P}_{t_k - t_{k-1}}(\mathbf{x}_{k-1}, d\mathbf{x}_k)f_k(\mathbf{x}_k) =$$
$$= \int \mu(d\mathbf{x}) \int \mathbf{P}_{t_1}(\mathbf{x}, d\mathbf{x}_1)f_1(\mathbf{x}_1)\tilde{h}(\mathbf{x}_1).$$

From (2) the above is equal to

$$\int \mu(d\mathbf{x})f_1(\mathbf{x})\tilde{h}(\mathbf{x}) \; .$$

Since f_1 and $\tilde{h}$ are nondecreasing, from association of μ we get

$$\mathbf{E}_\mu(f_1(\mathbf{X}(t_1)) \dots f_k(\mathbf{X}(t_k)) \geq \mathbf{E}_\mu f_1(\mathbf{X}(0))\mathbf{E}_\mu f_2(\mathbf{X}(t_2 - t_1) \dots f_k(\mathbf{X}(t_k - t_1)) \; ,$$

from the induction hypotheses this is greater than or equal to

$$\mathbf{E}_\mu f_1(\mathbf{X}(0))\mathbf{E}_\mu f_2(\mathbf{X}(t_2 - t_1)) \dots \mathbf{E}_\mu f_k(\mathbf{X}(t_k - t_1)) = \mathbf{E}_\mu f_1(\mathbf{X}(t_1)) \dots \mathbf{E}_\mu f_k(\mathbf{X}(t_k)).$$

The proof is complete. □

Let $\mathbf{X}^{\leftarrow} = (\mathbf{X}^{\leftarrow}(t) = \mathbf{X}(-t), t \in \mathbf{R})$ be the time reversal of $\mathbf{X}$, and let $\{\mathbf{T}^{\leftarrow}(t)\}$ denote the corresponding semi-group . For stochastically monotone Feller processes which have stochastically monotone time reversals we have a stronger positive time dependence (see Daduna and Szekli (1992)).

Theorem C. *Suppose that* $\mathbf{T}(t)f \in I^*_{(\mathcal{P})}$, *and* $\mathbf{T}^{\leftarrow}(t)f \in I^*_{(\mathcal{P})}$, *for each* $f \in I^*_{(\mathcal{P})}$, *and* $t \geq 0$, *i.e.* $\mathbf{X}$ *and* $\mathbf{X}^{\leftarrow}$ *are stochastically monotone. If* μ *is a stationary measure for* $\mathbf{X}$ *, which is associated*$(\prec_{(\mathcal{P})})$*, then*

$$\mathbf{E}_\mu\{f[\mathbf{X}(t_1),\ldots,\mathbf{X}(t_i)]g[\mathbf{X}(t_{i+1}),\ldots,\mathbf{X}(t_k)]\}$$

$$\geq \mathbf{E}_\mu\{f[\mathbf{X}(t_1),\ldots,\mathbf{X}(t_i)]\}\mathbf{E}_\mu\{g[\mathbf{X}(t_{i+1}),\ldots,\mathbf{X}(t_k)]\},$$

for nondecreasing f, g $, i = 1,\ldots,k,$ *and* $t_1 < \ldots < t_k, k \geq 1$.

Proof. Let $\mathbf{X}_1^i = (\mathbf{X}(t_1),\ldots,\mathbf{X}(t_i))$ and $\mathbf{X}_{i+1}^k = (\mathbf{X}(t_{i+1}),\ldots,\mathbf{X}(t_k))$. Then

$$\mathbf{E}_\mu(f(\mathbf{X}_1^i)g(\mathbf{X}_{i+1}^k)) = \int \mathbf{P}_\mu^{\mathbf{X}_1^i}(d\mathbf{x})f(\mathbf{x})\int \mathbf{P}_\mu^{(\mathbf{X}_{i+1}^k|\mathbf{X}_1^i=\mathbf{x})}(d\mathbf{y})g(\mathbf{y}) =$$

$$= \int \mathbf{P}_\mu^{\mathbf{X}_1^i}(d\mathbf{x})f(\mathbf{x})\int \mathbf{P}_\mu^{(\mathbf{X}_{i+1}^k|X(t_i)=x_i)}(d\mathbf{y})g(\mathbf{y})$$

$$= \int \mathbf{P}_\mu^{\mathbf{X}_1^i}(d\mathbf{x})f(\mathbf{x})h(x_i)$$

for some function h , which is nondecreasing from the assumption that $\mathbf{X}$ is stochastically monotone, where $\mathbf{P}_\mu^{\mathbf{X}}$ denotes the distribution of $\mathbf{X}$ when starting under μ, and the other expressions are interpreted similarly. Now we can write the last expression as

$$\int \mathbf{P}_\mu^{\mathbf{X}_1^i}(d\mathbf{x})f(\mathbf{x})h(x_i) = \int \mathbf{P}_\mu^{X(t_i)}(dx_i)\int \mathbf{P}_\mu^{(\mathbf{X}_1^{i-1}|X(t_i)=x_i)}(d(x_1,\ldots,x_{i-1}))f(\mathbf{x})h(x_i)$$

$$= \int \mathbf{P}_\mu^{X(t_i)}(dx_i)\tilde{f}(x_i)h(x_i),$$

for $\tilde{f}$ which is nondecreasing from the assumption that $\mathbf{X}^{\leftarrow}$ is stochastically monotone. In equilibrium $\mathbf{P}_\mu^{X(t_i)} = \mu$ and from the assumption that μ is associated-$(\prec_{(\mathcal{P})})$ we have

$$\int \mathbf{P}_\mu^{X(t_i)}(dx_i)\tilde{f}(x_i)h(x_i) \geq \int \mathbf{P}_\mu^{X(t_i)}(dx_i)\tilde{f}(x_i)\int \mathbf{P}_\mu^{X(t_i)}(dx_i)h(x_i).$$

Writing $\tilde{f}$ and h in this expression we see that the last term is equal to

$$\mathbf{E}_\mu\{f[\mathbf{X}(t_1),\ldots,\mathbf{X}(t_i)]\}\mathbf{E}_\mu\{g[\mathbf{X}(t_{i+1}),\ldots,\mathbf{X}(t_k)]\},$$

which completes the proof. □

In general, monotonicity of $\mathbf{X}$ does not imply monotonicity of $\mathbf{X}^{\leftarrow}$, so the conditions in Theorem C. are not dispensable.

3.8 Dependencies in Markovian networks

For a queueing network which is described by the joint queue lengths process $\{\mathbf{X}(t), t \geq 0\}$ several dependence structures are of interest. For example dependence of the number of customers at different nodes, for a fixed time, i.e. $X_i(t), X_j(t)$, or for more than one fixed times. Taking functionals of the process $\{\mathbf{X}(t), t \geq 0\}$ one can define many other interesting characteristics, for example sojourn times of customers in specified nodes or on specified routes; inter-departure times, inter-output times etc. Such dependencies are of interest for queues in equilibrium, as well as in their transient state. The strength of dependence may be measured potentially with a use of one from a lot of existing dependence notions, however in the literature most of the results are for weakest dependencies measured by covariance . A few results are known for stronger dependence notions, e.g. association, and their proofs indicate that obtaining them was not a simple task.

The most extensive work directed to the task of systematizing and recognizing dependence structures in queueing networks, especially for traffic processes is included in the book by Disney and Kiessler (1987) .
For time dependent processes $\{Y(t)\}, \{Z(t)\}$ define cross correlation of lag r by

$$\text{ccorr}\ (Y(t), Z(t+r)) = \text{ Cov }(Y(t), Z(t+r))/(\text{Var } Y(t) \text{ Var } Z(t+r))^{1/2},$$

and similarly auto covariance and auto correlation of lag r , for a single time dependent process.
For a delayed feedback queue denote by $X_1(t)$ the number of customers in the basic queue. Disney and Kiessler (1987) studied by simulation for example the value of $\text{ccorr}(X_1(t), X_1(t+r))$ in steady state, as a function of r, when both queue capacities are infinite, and the queue disciplines are FCFS. The obtained value was positive for $r \neq 0$.

They also studied by simulation traffic processes in three-node cyclic networks, delayed feedback closed networks and some other networks. No negative auto correlations were found by simulation in open exponential Jackson networks , and Disney and Kiessler (1987, p.62) conjectured that this may always be true.

Further facts related to the problem of dependence structures in networks and reported in Disney and Kiessler (1987) are the following:

- For $M/M/1/0$ queue

$$\lim_{t\to\infty} \text{ccorr}\ (N^{ov}(t), N^{i}(t)) = (\lambda\mu^2 - \lambda^2\mu)/[(\lambda^2 + 4\lambda\mu + \mu^2)(\lambda\mu^3 + \lambda^3\mu)]^{1/2},$$

  where $N^{ov}(N^{i})$ denotes the number of overflows (inputs) in $(0, t]$ (λ, μ are arrival and service rates, respectively).

- Covariance (lag1) in $M/GI/1/L, 0 < L \leq \infty$, output processes:

  Suppose that $D_1^o, D_2^o, \ldots$ denote the consecutive output intervals. Then

$$\text{Cov}\ (D_m^o, D_{m+1}^o) = v^0(0)[p' + p^2/\lambda - (ES + v^0(0)/\lambda)]/\lambda p,$$

  where v^0 is the stationary distribution of the queue length embedded at output epochs, and

$$p = \int_0^\infty e^{-\lambda y} H(dy),$$

$$p' = \int_0^\infty -ye^{-\lambda y} H(dy),$$

$$ES = \int_0^\infty yH(dy),$$

for the service time distribution H and the arrival rate λ. In particular, for $M/M/1/L, L \geq 1$,and $M/E_k/1/1$

$$\text{Cov}\ (D^\circ_m, D^\circ_{m+1}) \leq 0.$$

Further results related to the problems of dependence in networks are as follows

- A positive dependence for the number of customers at different nodes ,and at different times, when the system is in equilibrium is known for the $(./GI/\infty)^n$ Jackson network (see Kanter (1985)). Let

$$B_j = \{X_i(t_j) \leq k_{ij}; i \in \{1, \ldots, n\}\},$$

for $j = 1, \ldots, k, t_1 < \ldots < t_k, k_{ij} \in \mathbf{N}$ (n denotes the number of nodes in the network) . Then

$$\mathbf{P}_\pi(\cap_{j=1}^k B_j) \geq \pi(B_1) \ldots \pi(B_k),$$

where π denotes the stationary distribution for $(\mathbf{X}(t), t \geq 0)$, and $\mathbf{P}_\pi$ is the process distribution under initial distribution π

- Consider now the following Jackson network. Customers arrive according to a Poisson process with rate λ . All customers enter node 1. Service times at nodes are mutually independent and exponentially distributed. Customers choose either the route $r_1 \equiv 1 \to 2 \to 3$ or $r_2 \equiv 1 \to 3$ according to a Bernoulli process with parameter p of choosing r_1 . This decision process, the arrival, and the service processes are mutually independent . Let S_1 and S_3 be the sojourn times at nodes 1 and 3 of a customer that follows route r_1 . Foley and Kiessler (1989) showed that S_1 and S_3 are associated.

- Recently McNickle (1991) considered lagged queue length correlations in a two-node Jackson network. He proved that

$$\text{Cov}\ (X_1(0), X_2(t)) / \ \text{Cov}\ (X_2(0), X_1(t)) = \gamma_1 r_{12}/\gamma_2 r_{21},$$

where $R = \{r_{ij}\}$ is the routing matrix and γ_1, γ_2 are the rates of flow of customers through the two nodes. Now if we have been able to measure or calculate the lagged correlation between node 1 and node 2, we also know what it is between node 2 and node 1.

A stochastic processes approach gives stronger results.

Theorem A. *Suppose that* $\mathbf{X}$ *is a* $\leq_{st}$*-monotone generalized birth-death process with an initial distribution* μ *which is associated. Then* $\mathbf{X}$ *is associated in time.*

Proof. This is a direct consequence of Theorem 3.7 A. since this process is of "up-down" type in the sense of Harris (1977) with respect to the coordinatewise ordering $\leq$ and Liggett's condition is easily verifiable (see Liggett (1985) , Chapter II, Theorem 2.13 and the discussion there) . □

The "up-down" property of jumps used in the proof of the above theorem is crucial in the sense that positive correlations of monotone processes are equivalent to the fact that direct jumps between states which are not comparable under $\leq$ are not allowed. Unfortunately in the systems of our main interest- queueing networks - this property fails to hold.

However, Jackson networks possess some positive dependence properties "over time", which hold in equilibrium. This is a consequence of Theorem 3.7.C. and the known fact that the time reversals of Jackson networks are again Jackson networks, and as such are stochastically monotone (see e.g. Kelly (1979), pp.51, 61).

Theorem B. (Daduna and Szekli (1992)) *If* $\mathbf{X}$ *is a Jackson network with non-decreasing service rates in equilibrium with stationary measure* μ*, then*

$$\mathbf{E}_\mu\{f[\mathbf{X}(t_1),\ldots,\mathbf{X}(t_i)]g[\mathbf{X}(t_{i+1}),\ldots,\mathbf{X}(t_k)]\}$$

$$\geq \mathbf{E}_\mu\{f[\mathbf{X}(t_1),\ldots,\mathbf{X}(t_i)]\}\mathbf{E}_\mu\{g[\mathbf{X}(t_{i+1}),\ldots,\mathbf{X}(t_k)]\}$$

for all non-decreasing real f*,* g*, and* $0 \leq t_1 < \ldots < t_k, i < k, i, k \in \mathbf{N}$.

From the above inequality every pair $(X_i(t), X_j(s))$ from stationary Jackson network, for $0 \leq s < t$ and $i, j \in \{1,\ldots,n\}$, is positively quadrant dependent (PQD) i.e.

$$P(X_i(t) > (\leq)u, X_j(t) > (\leq)v) \geq P(X_i(t) > (\leq)u)P(X_j(t) > (\leq)v).$$

for all $u, v \in R_+$ (see Lehmann (1966)).

The functions f, g can be taken both non-increasing, so approximating appropriate indicator functions by monotone f's and g's, and using induction one obtains a generalization of Kanter's (1985) result for networks of $./M/\infty$-queues on overload probabilities to arbitrary Jackson networks with service rates non-decreasing in queue length.

Theorem C. *If* $\mathbf{X}$ *is a Jackson network in equilibrium with stationary measure* μ*, then*

$$\mathbf{P}_\mu(\cap_{i=1}^k (X_j(t_i)) \leq K_{ij}, j = 1,\ldots,n) \geq \prod_{i=1}^k \mathbf{P}_\mu(X_j(t_i) \leq K_{ij}, j = 1,\ldots,n)$$

where $K_{ij} \in \mathbf{N}, 0 \leq t_1 < \ldots < t_k, k \in \mathbf{N}$.

Note that the inequalities in brackets in the above inequality can be reversed.

The following result is a direct consequence of Theorem 3.7.A. (see Section 2.12 for definition of b-d-m process).

Theorem D. *If* $\mathbf{X}$ *is a* $<_{st*}$*-monotone simple* $b-d-m$ *process, with an initial distribution* μ*, which is associated-*$(\leq_*)$*, then* $\mathbf{X}$ *is associated-*$(\leq_*)$ *in time.*

It is remarkable that the use of the "up-down" property in the above theorem does not depend on how the nodes of the network are numbered. It should be mentioned also that neither stationarity nor product form requirements are of importance here.

The following example combines the results of Theorems A. and D., by using the concept of **partition separated orders** as in Section 3.7.

Example. (See Figure 1)

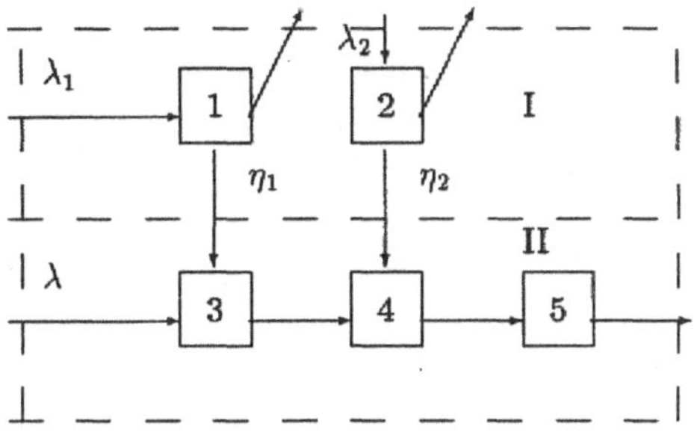

Figure 1

We consider a network of nodes $\tilde{J} = \{1, 2, ..., 5\}$, which act as $./M/1/\infty$ FCFS nodes

with state dependent service rates $\mu_j(.) > 0$, which are non-decreasing in the queue length ($\mu_j(0) = 0, j = 1, ..., 5$). Nodes 3,4,5 are fed by a Poisson(λ) stream of jobs which is independent of all service times. All the service times constitute an independent family. Nodes 1 and 2 are independent $M/M/1/\infty$ FCFS systems serving each a stream of customers arriving in a Poisson(λ_i) process , i=1,2. After being served these customers leave the network immediately. As long as nodes 1,2, which constitute *subnetwork I*, are busy they produce an additional load for *subnetwork II*, which consists of nodes 3,4,5. As node 1 (2, resp.) is working with rate η_1 (η_2, resp.) , it produces jobs entering node 3 (4, resp.) independently of anything else. Note that sending out a job for node 3 (4, resp.) does not change the state of node 1 (2, resp.). We denote by $\mathbf{X} = (\mathbf{X}(t), t \geq 0)$, $\mathbf{X}(t) = (X_j(t), j \in \tilde{J})$ the Markovian joint queue length process of the nodes with the state space $E = N^5$.
It is easy to show that $\mathbf{X}$ is monotone under the coordinatewise ordering in which each coordinate carries the natural order on $\mathbf{N}$. But the process does not fulfill the criterion of Theorem 3.7.A., which in the case of discrete state space is just the *up-down* property. The latter is not given for movements of customers from the node 3 to the node 4. On the other hand, under $\leq_*$ the network process is not monotone (this does not depend on the numbering of the nodes). However, we can partition our node set $\tilde{J}$ as follows. Set $\mathcal{P} = \{(1,2),(3,4,5)\}$ and assume on the node set $(1,2)$ the coordinatewise ordering under the natural order on N, and on the node set $(3,4,5)$ the partial sum order on $\mathbf{N}^3$, where the nodes are ordered according to their numbering. From Section 2.8 we know: $\mathbf{X}$ is $\prec_{st\mathcal{P}}$ monotone, where $\prec_{st\mathcal{P}}$ is the strong stochastic ordering generated by the order above. We have that all transitions of $\mathbf{X}$ are either *up* or *down*, so Theorem 3.7.A. applies and $\mathbf{X}$ is associated ($\prec_{st\mathcal{P}}$) in time.

The above example is a prototype of a coupling of subnetworks, each having its own order structure.

Now we turn our attention to closed networks.

Theorem E. *Let* $(\mathbf{X}(t), t \geq 0)$ *denote the joint queue length process of an irreducible Gordon-Newell network with Markovian routing and queue-length dependent increasing service rates, which acts in equilibrium. Then for each* $t \geq 0$, $\mathbf{X}(t)$ *is negatively associated (NA).*

Proof. We may apply Theorem 3.5.H. This is possible because the distribution of $\mathbf{X}(t)$ is the conditional steady state joint queue length distribution of a suitable open Jackson network, which has independent coordinates for fixed time point t. We summarize the needed well-known facts as follows:
Given any open Jackson network with nodes $1, \ldots, n$, irreducible Markovian routing and state-dependent service rates $\mu_i(k), k \in \mathbf{N}, i = 1, \ldots, n$, which are non-decreasing in k, let $\mathbf{X} = (X_1, \ldots, X_n)$ be a vector, distributed according to the equilibrium distribution of the network. Then
1) For any $A \subseteq \{1, \ldots, n\}, A \neq \emptyset$, there exists an ergodic Jackson network with irreducible Markovian routing having stationary distribution $\mathbf{P}^{(X_j, j \in A)}$.
2) For any $A \subseteq \{1, \ldots, n\}, A \neq \emptyset$, and any $k > 0$, there exists an ergodic Gordon-Newell network with irreducible Markovian routing, and k customers cycling in the network, such that the stationary distribution of the network is given by

$$\mathbf{P}^{(X_j, j \in A | \sum_{j \in A} X_j = k)},$$

(see Gordon and Newell (1967)).
In view of the above representation, to apply Theorem 3.5.H. we have to show that for any Gordon-Newell network the following holds:
If $r : \mathbf{N}^n \to \mathbf{R}$ is nonnegative and increasing with respect to the coordinatewise ordering, and if $\mathbf{Y} = (Y_1, \ldots, Y_n)$ is distributed according to the stationary distribution in the network then $\mathbf{E}[r(\mathbf{Y})]$ is increasing in the population size of the network. But this result is known, e.g. in Van der Wal (1989). □

Remark. From Theorem 3.5.E. we may deduce another negative dependency property of Gordon-Newell networks with increasing service rates, namely the joint queue length in any of J-1 of J nodes is RR_2 in pairs (see Section 3.5 for definition).

3.9 Dependencies in Markov renewal queues

We obtain some dependence properties for the MR arrival process $\mathbf{D}$, represented by $[\boldsymbol{\pi}, \mathbf{A}, \mathbf{F}]$.

Theorem A. *If* $F_k <_{st} F_{k+1}$ *for all* $k \geq 1$, *and* $\mathbf{Z}$ *is stochastically monotone then for each* $n = 1, 2, \ldots,$ $\mathbf{D}\mid_n = (D_1, \ldots, D_n)$ *is associated.*

Proof. . Suppose that $f, g : R^n \to R$ are nondecreasing. We have

$$\begin{aligned}
\int_\Omega f(\mathbf{D}\,|_n)g(\mathbf{D}\,|_n)dP &= \int_\Omega f(X^{(1)}_{Z_1},\ldots,X^{(n)}_{Z_n})g(X^{(1)}_{Z_1},\ldots,X^{(n)}_{Z_n})dP \\
&= \int\int f(x^{(1)}_{z_1},\ldots,x^{(n)}_{z_n})g(x^{(1)}_{z_1},\ldots,x^{(n)}_{z_n})dP^{\mathbf{X}}(\mathbf{x})dP^{\mathbf{Z}}(\mathbf{z}) \\
&\geq \int[\int f(x^{(1)}_{z_1},\ldots,x^{(n)}_{z_n})dP^{\mathbf{X}}(\mathbf{x})\int g(x^{(1)}_{z_1},\ldots,x^{(n)}_{z_n})dP^{\mathbf{X}}(\mathbf{x})]dP^{\mathbf{Z}}(\mathbf{z}) \\
&= \int \Phi(\mathbf{z})\Gamma(\mathbf{z})dP^{\mathbf{Z}}(\mathbf{z}),
\end{aligned}$$

where $\mathbf{z} = (z_1, \ldots, z_n)$ and $\mathbf{x} = \{x^{(j)}_i,\ i \in \mathcal{S},\ j = 1, \ldots, n\}$. In the above the second equality is from independence of $\mathbf{Z}$ and $\mathbf{X}$, and the inequality is from the fact that vectors of independent random variables are associated. Note that the above functions $\Phi(\mathbf{z})$ and $\Gamma(\mathbf{z})$ are nondecreasing in $\mathbf{z}$. This follows from the assumption that $F_k <_{st} F_{k+1}$, and the definition of stochastic ordering. From this and the fact that $\mathbf{Z}$ is stochastically monotone and ; hence associated (see section 2), we obtain

$$\int \Phi(\mathbf{z})\Gamma(\mathbf{z})dP^{\mathbf{Z}}(\mathbf{z}) \geq \int \Phi(\mathbf{z})dP^{\mathbf{Z}}(\mathbf{z})\int \Gamma(\mathbf{z})dP^{\mathbf{Z}}(\mathbf{z}).$$

Finally we have

$$\int_\Omega f(\mathbf{D}\,|_n)g(\mathbf{D}\,|_n)dP \geq \int_\Omega f(\mathbf{D}\,|_n)dP \int_\Omega g(\mathbf{D}\,|_n)dP,$$

which means that $\mathbf{D}\,|_n$ is associated. □

Under the assumption that $F_{k+1} <_{st} F_k$, for $k = 1, 2, \ldots$ (and keeping other assumptions without change), we have the same result that $\mathbf{D}\,|_n$ is associated. This follows from the fact that in the definition of association we can take two nonincreasing functions instead of two nondecreasing ones. Note that stochastic monotonicity of $\mathbf{Z}$ depends only on the matrix $\mathbf{A}$, not on initial conditions.

Example. (Patuwo et al. (1991)) Consider a Markov renewal arrival process with the two dimensional matrix

$$\mathbf{A} = \begin{bmatrix} a & 1-a \\ 1-b & b \end{bmatrix}$$

The corresponding semi-Markov kernel is of the form

$$\mathbf{A}(t) = \begin{bmatrix} aF_1(t) & (1-a)F_2(t) \\ (1-b)F_1(t) & bF_2(t) \end{bmatrix}$$

We immediately see that $\mathbf{A}$ is stochastically monotone if $\xi = a + b - 1 \geq 0$. Hence every choice of distributions F_1 and F_2, which are stochastically ordered gives an associated arrival stream $\mathbf{D}\,|_n$ if $\xi \geq 0$. In Patuwo (1991), F_1 and F_2 are Erlang with the same number of phases and with different means, thus they are stochastically ordered. Positively correlated arrival processes considered in this paper are associated.

It is of some independent interest to know that the dependence in the arrival process can be stronger than association.

Lemma B. *If $F_k <_{lr} F_{k+1}$, $k = 1, 2, \ldots$ and $\mathbf{A}$ is a TP_2 matrix then for each $n = 1, \ldots, \mathbf{D}\mid_n$ is a MTP_2 vector.*

Example. (continuation). From the definition of TP_2 matrices it is clear that the two dimensional matrix $\mathbf{A}$ in Patuwo et al. (1991) is TP_2 if $a + b - 1 \geq 0$. The Erlang distributions from this paper have the same number of stages, so if $\lambda_1 \geq \lambda_2$ it is immediate from the definition that $F_1 <_{lr} F_2$. Thus positive correlations studied in this paper are for $\lambda_1 \geq \lambda_2$ of type MTP_2.

We can generalize the above results to the case when the holding times in the MR arrival process depend on both the actual and the previous state value.
Consider a sequence of arrays of independent, positive random variables $\mathbf{X} = \{X_{ij}^{(n)}, i, j, n = 1, 2, \ldots\}$. For each fixed i, j the sequence $\{X_{ij}^{(n)}, n = 1, 2, \ldots\}$ is assumed to be i.i.d. with a common distribution F_{ij} . Now the arrival process with interpoint distances $D_n = X_{Z_{n-1},Z_n}^{(n)}$, $n = 1, 2, \ldots$, is a general MR arrival process with the corresponding semi-Markov kernel $\mathbf{A}(t) = \{a_{ij}F_{ij}(t)\}$. Let $\mathbf{F} = \{F_{ij}\}$.

Theorem C. *If $F_{ij} <_{st} F_{i'j'}$ for all $i \leq i'$, and $j \leq j'$, and $\mathbf{Z}$ is stochastically monotone, then for each $n = 1, 2, \ldots, \mathbf{D}\mid_n = (D_1, \ldots, D_n)$ is associated.*

Remark. Under the assumption that $F_{ij} >_{st} F_{i',j'}$, for $i \leq i'$, $j \leq j'$ (and keeping other assumptions without change), we have the same result that $\mathbf{D}\mid_n$ is associated. This follows from the fact that in the definition of association we can take two nonincreasing functions instead of two nondecreasing ones. Note that stochastic monotonicity of $\mathbf{Z}$ depends only on the matrix $\mathbf{A}$, not on initial conditions.

Theorem D. *If $F_{ij} <_{lr} F_{i',j'}$, $i \leq i'$, $j \leq j'$, and $\mathbf{A}$ is a TP_2 matrix then for each $n = 1, \ldots$, $\mathbf{D}\mid_n$ is a MTP_2 vector.*

Our interest now is to find the influence of the correlation structure of the arrival process on the queue under changes of the governing matrix.
Let us consider, for $n = 2, 3, \cdots$, an n-state MR arrival process of which $\mathbf{A} = \mathbf{A}_n(p)$

$$\mathbf{A}_n(p) = \begin{pmatrix} p & \frac{1-p}{n-1} & \cdots & \frac{1-p}{n-1} \\ \frac{1-p}{n-1} & p & \cdots & \frac{1-p}{n-1} \\ \vdots & \vdots & \ddots & \vdots \\ \frac{1-p}{n-1} & \frac{1-p}{n-1} & \cdots & p \end{pmatrix}.$$

We choose matrices of the above form to keep the one dimensional distributions of the arrival process fixed after a change of the governing matrix. he $\mathbf{A}_n(p)$ matrix is reversible, so it gives a reversible arrival process (see e.g., Disney and Kiessler (1987)). For the stationary, ergodic MR arrival process represented by $[\boldsymbol{\pi}, \mathbf{A}_n(p), \mathbf{F}]$, the one-dimensional marginal distributions $P(D_i \leq t)_p$, $i = 1, 2, \ldots$ are independent of p since $\boldsymbol{\pi} = \frac{1}{n}e^T$, $(e^T = (1, \ldots, 1))$, and

$$P(D_i \leq t)_p = \boldsymbol{\pi}\mathbf{A}_p(t)e = \frac{1}{n}\sum_{i=1}^{n} F_i(t).$$

where $\mathbf{A}_p(t)$ corresponds to $\mathbf{A}_n(p)$.

Remark. Not only for $\mathbf{A}_n(p)$, but also for any doubly stochastic matrix, the distribution $P(D_i \le t)$ does not depend on the matrix, because in this case $\boldsymbol{\pi} = \frac{1}{n}e^T$. However, if we require in addition reversibility, a natural choice is to take $\mathbf{A}_n(p)$.

An expression for the correlation structure in the arrival process $[\boldsymbol{\pi}, \mathbf{A}_n(p), \mathbf{F}]$ is given in the following theorem.

Theorem E. *Let for* $[\boldsymbol{\pi}, \mathbf{A}_n(p), \mathbf{F}]$

$$\int_R x dF_i(x) = m_i \qquad \int_R (x - m_i)^2 dF_i(x) = v_i$$

be finite for $i = 1, 2, \cdots, n$. *Then*

$$\begin{aligned} Corr(r) &= Corr(D_k, D_{k+r}) \\ &= \frac{\frac{1}{n^2}\sum_{i<j}(m_i - m_j)^2 \xi_n{}^r}{\frac{1}{n}\sum_{i=1}^n v_i + \frac{1}{n^2}\sum_{i<j}(m_i - m_j)^2}, \end{aligned}$$

where $\xi_n = \frac{np-1}{n-1}$.

Proof. Let $k, r \ge 1$. We have from definition

$$\begin{aligned} Corr(r) &= Corr(D_k, D_{k+r}) \\ &= \frac{E(D_k, D_{k+r}) - (E(D_k))^2}{Var(D_k)}. \end{aligned}$$

For notational convenience, let $\mathbf{A} \equiv \mathbf{A}_n(p)$. From Patuwo (1989), we have

$$\begin{aligned} E(D_k) &= \boldsymbol{\pi}\mathbf{A}^{(1)}e = \frac{1}{n}\sum_{i=1}^n m_i \\ Var(D_k) &= E(D_k{}^2) - (E(D_k))^2 \\ &= \frac{1}{n}\sum_{i=1}^n v_i + \frac{1}{n^2}((n-1)\sum_{i=1}^n m_i{}^2 - \sum_{i \ne j} m_i m_j) \\ E(D_k D_{k+r}) &= \boldsymbol{\pi}\mathbf{A}^{(1)}\mathbf{A}^{r-1}\mathbf{A}^{(1)}e \end{aligned}$$

We can compute $\mathbf{A}^{r-1}$ by using a similarity transform, i.e., $\mathbf{A}^l = \mathbf{N}\mathbf{D}^l\mathbf{N}^{-1}$, $l = 1, 2, \cdots$, where $\mathbf{D}$ is an $n \times n$ diagonal matrix of eigenvalues of $\mathbf{A}$, and $\mathbf{N}$ is a nonsingular matrix with columns being eigenvectors of $\mathbf{A}$. It is not difficult to get the eigenvalues of $\mathbf{A}$, λ_i, $i = 1, \cdots, n$, which are $\lambda_1 = 1$, $\lambda_2 = \cdots = \lambda_n = \frac{np-1}{n-1}$ and the eigenvectors of $\mathbf{A}$, f_i, $i = 1, \cdots, n$, corresponding to λ_i, which are $f_1 = (1, \cdots, 1)^T$, $f_2 = (1, -1, 0, \cdots, 0)^T$, $f_3 = (1, 0, -1, \cdots, 0)^T$, $\cdots$, $f_n = (1, 0, \cdots, -1)^T$.
With some algebra, we get

$$\mathbf{A}^l = \mathbf{N}\mathbf{D}^l\mathbf{N}^{-1} = e\boldsymbol{\pi} + \xi_n{}^l(\mathbf{I} - e\boldsymbol{\pi}),$$

where $\xi_n = \frac{np-1}{n-1}$. So,

$$\begin{aligned} E(D_k D_{k+1}) &= \pi \mathbf{A}^{(1)}(e\pi + \xi_n{}^{r-1}(\mathbf{I} - e\pi))\mathbf{A}^{(1)}e \\ &= (E(D_k))^2 + \xi_n{}^{r-1}(\pi \mathbf{A}^{(1)}\mathbf{A}^{(1)}e - (E(D_k))^2). \end{aligned}$$

Finally we get

$$Corr(r) = \frac{\frac{1}{n^2}\sum_{i<j}(m_i - m_j)^2 \xi_n{}^r}{\frac{1}{n}\sum_{i=1}^n v_i + \frac{1}{n^2}\sum_{i<j}(m_i - m_j)^2},$$

which completes the proof. □

$Corr(r) \geq 0$ if and only if $\xi_n \geq 0$, that is, $p \geq \frac{1}{n}$. Thus, $p \geq \frac{1}{n}$ is sufficient and necessary condition for positive correlations.

We restrict now our attention to MR/M/1 queues with the arrival stream $[\pi, \mathbf{A}_2(p), \mathbf{F}]$. This two dimensional setting allows us to find a formula for expectation of the stationary queue length embedded at the arrival times. The implications we can draw from this formula confirm the intuitive understanding that the mean queue length can in some situations rapidly increase when correlations in the arrival process increase (the traffic intensity is kept constant). In many numerical studies it was found that the factor of growth can be greater than forty (for details see Patuwo et al. (1991)). In fact, under some assumptions on m_i's , the mean queue length tends to infinity with p tending to 1, so from Theorem E. we see that the increase of the correlations in the arrival process through the variable p can be dramatic, and the factor of growth can be arbitrary large. However, it is important for this result that the correlations are increased through p, we can increase the correlations alternatively by reducing variability (and the variation) of F_j's , but in this case the mean queue length will be smaller, because the arrival stream will be less variable.

Let the semi-Markov kernel be

$$\mathbf{A}(t) = \begin{bmatrix} pF_1(t) & (1-p)F_2(t) \\ (1-p)F_1(t) & pF_2(t) \end{bmatrix},$$

where $F_i(t) = 1 - e^{-\lambda_i t}$, $i = 1, 2$, $t \geq 0$. Suppose that the service time is exponentially distributed with the rate μ. The "individual" traffic intensities are $\rho_1 = \lambda_1/\mu$, and $\rho_2 = \lambda_2/\mu$. The traffic intensity for the system is the harmonic mean of the individual traffic intensities $\rho = \{\frac{1}{2}(\frac{1}{\rho_1} + \frac{1}{\rho_2})\}^{-1}$. We assume that $\rho < 1$.

Denote by N_n^a the number of customers in the system embedded just before the n-th arrival epoch.

Theorem F. (Szekli et al. (1993b)) *For a stationary MR/M/1 queue with the arrival process $[\pi, \mathbf{A}_2(p), \mathbf{F}]$ and $\rho < 1$*

$$EN^a = \frac{\rho}{1-\rho} + \frac{1 - \frac{2(1-P_0)}{\rho_1 + \rho_2}}{2(1-p)(1-\rho)}. \qquad (1)$$

From the above result we have the following important theorem.

Theorem G. *Consider a stationary MR/M/1 queue with the arrival process* $[\boldsymbol{\pi}, \mathbf{A}_2(p), \mathbf{F}]$ *and* $\rho < 1$. *If the arithmetic mean* $(\rho_1 + \rho_2)/2 > 1$ *then*

$$EN^a \to \infty, \text{ as } p \to 1.$$

Proof. Looking at the second term in the right-hand side of (1) we see that it is enough to show that

$$1 - \frac{2(1-P_0)}{\rho_1 + \rho_2} \geq \epsilon,$$

for some $\epsilon > 0$.
However this inequality is equivalent to

$$P_0 \geq 1 - \frac{\rho_1 + \rho_2}{2} + \epsilon \rho_1 \rho_2,$$

which is fulfilled if

$$\frac{\rho_1 + \rho_2}{2} \geq 1 + \epsilon \rho_1 \rho_2.$$

It is clear that the last inequality follows from the assumption that $\frac{\rho_1+\rho_2}{2} > 1$, for some $\epsilon > 0$. □

3.10 Associated point processes

There are two ways to define association of point processes. The first approach is based on the classical notion of association of random variables introduced by Esary et al.(1967). From this viewpoint, a point process N is a collection of random variables $\{N(B) : B \in \mathcal{B}\}$, where $\mathcal{B}$ is a σ-ring of Borel sets. The point process N is then **associated** iff $(N(B_1), \ldots, N(B_n))$ is a vector of associated random variables for any $n \geq 1$ and $B_1, \ldots, B_n \in \mathcal{C}$, where $\mathcal{C}$ is a certain subset of $\mathcal{B}$.

The other approach is to think of N as a random element assuming its values in some Polish space endowed with a closed partial order $<$. A point process N is then said to be associated if and only if $\text{Cov}(f(M), g(M)) \geq 0$ for any pair of real-valued functions f, g on $\mathcal{S}$, nondecreasing with respect to the order $<$. See Lindqvist (1988) for more detailed treatment.

Let $\mathcal{N}$ be a space of Radon (i.e., locally finite) integer valued measures on a space E. Let $\mathcal{F}_c = \mathcal{F}_c(E)$ be a class of nonnegative continuous functions $E \to R$ with compact supports. The space $\mathcal{N}$ can be endowed with the *vague* topology with base files of the form $\{\mu \in \mathcal{N} : s < \int_E f\, d\mu < t\}$ for $s, t \in R$ and $f \in \mathcal{F}_c$ (see Kallenberg (1983)). Define the following partial order $<_{\mathcal{N}}$ in $\mathcal{N}$. For $\mu_1, \mu_2 \in \mathcal{M}$ let $\mu_1 < \mu_2$ if and only if $\mu_1(B) \leq \mu_2(B)$ for all (topologically) bounded sets $B \in \mathcal{B}(E)$.

The space $\mathcal{N}$ is metrizable as a Polish space. Moreover, the order $<_{\mathcal{N}}$ is closed in $\mathcal{N}$ (Kallenberg (1983)). Thus, $\mathcal{N}$ satisfies all assumptions in Lindqvist (1988) necessary to define the association of an $\mathcal{N}$-valued random element N, i.e., a point process N on E. And so from now on, we shall say that a point process N is associated if and only if

$$\text{Cov}(f(N), g(N)) \geq 0 \ ,$$

for all bounded increasing measurable functions $f, g : \mathcal{N} \to R$.

Rolski and Szekli (1991) considered a mapping $\gamma : \mathcal{N} \to R_+^\infty$ defined as follows. Let $\mathcal{I} = \{I_1, I_2, \ldots\}$ be a denumerable DC-semiring generating the ring of all bounded Borel sets in E (Kallenberg (1983)). For $\mu \in \mathcal{N}$ let,

$$\gamma(\mu) = (\mu(I_1), \mu(I_2), \ldots) .$$

Let $\mathcal{G} = \gamma(\mathcal{N})$. Lemma 2 in Rolski and Szekli (1991) yields the following lemma.

Lemma A. *$\mathcal{G}$ is a closed subset of R_+^∞, and $\gamma : \mathcal{N} \to \mathcal{G}$ is a homeomorphism, where the topology in $\mathcal{G}$ is the one generated by the topology in R_+^∞.*

The following theorem is a consequence of Lemma A..

Theorem B. *A point process N is associated iff random vectors $(N(B_1)), \ldots, N(B_n))$ are associated for all $n \geq 1$ and bounded sets $B_1, \ldots, B_n \in \mathcal{B}(E)$.*

Proof. Consider the natural coordinate order $\leq$ in R_+^∞. Both γ and γ^{-1} are increasing with respect to $<$ and $\leq$ respectively. Therefore, N is associated if and only if $\gamma(N)$ is associated as a random element of R_+^∞ (Lindqvist (1988)). But $\gamma(N)$ is associated if and only if $(N(I_1), \ldots, N(I_n))$ is a vector of associated random variables for all $n \geq 1$ (Lindqvist (1988), Theorem 5.1). Since this is true for any DC-semiring $\mathcal{I} = \{I_1, I_2, \ldots\}$ as specified above, the statement of the theorem follows. □

By an approximation argument, bounded Borel sets in the statement of Theorem B. can be replaced with a) all Borel sets for which $N(B_i) < +\infty$ a.s., b) all Borel sets, if one allows random variables to take values in the extended real space (with $-\infty$ and $+\infty$ as the extreme points) and defines association of random vectors so as to take that into account.

We shall assume that $E = R_+$. Let

$$\mathcal{J}_+ = \{(x_1, x_2, \ldots) \in (R_+ \cup \{+\infty\})^\infty : x_1 \leq x_2 \leq \ldots,\ x_n \to +\infty\} .$$

An element of $\mathcal{J}_+$ will be denoted by $\boldsymbol{t} = (t_1, t_2, \ldots)$. Consider $\mathcal{J}_+$ to be endowed with the usual product topology. For $\mu \in \mathcal{N}(R_+)$ let $\tau_0(\mu) = 0$ and for $n \geq 1$ let,

$$\tau_n(\mu) = \sup\{u > 0 : \mu(0, u] \leq n\} .$$

Define a mapping $\boldsymbol{\tau} : \mathcal{N}(R_+) \to \mathcal{J}_+$ as,

$$\boldsymbol{\tau}(\mu) = (\tau_1(\mu), \tau_2(\mu), \ldots) .$$

Note that $\boldsymbol{\tau}$ is invertible and that $\boldsymbol{\tau}^{-1}(\boldsymbol{t})$ is defined as a μ such that $\mu(A) = \sum_{n=1}^\infty \mathbf{1}_A(t_n)$, for all measurable sets $A \subset R_+$. Here, $\mathbf{1}_A(\cdot)$ is the indicator function of a set A.

Lemma C. *The mappings $\boldsymbol{\tau}$ and $\boldsymbol{\tau}^{-1}$ are measurable.*

Let $N : (\Omega, \mathcal{F}, P) \to \mathcal{N}(R_+)$, where $(\Omega, \mathcal{F}, P)$ is a probability space, be a point process on R_+. For brevity we shall write T_n for $\tau_n(N)$, $n \geq 0$. Note that N is uniquely determined by $(T_1, T_2, \ldots)$. Suppose that $T_n < T_{n+1} < +\infty$ a.s. for all $n \geq 1$.

Suppose that, for $n \geq 0$, regular versions of conditional probability distributions

$$F_{n+1}(t; t_1, \ldots, t_n) = P\{T_{n+1} \leq t \mid T_1 = t_1, \ldots, T_n = t_n\} ,$$

are continuous with respect to the Lebesgue measure on R_+. Denote the corresponding probability density functions by $f_{n+1}(t; t_1, \ldots, t_n)$. Let $r_{n+1}(t; t_1, \ldots, t_n)$ be the conditional failure rate of T_{n+1}. That is,

$$r_{n+1}(t; t_1, \ldots, t_n) = \frac{f_{n+1}(t; t_1, \ldots, t_n)}{1 - F_{n+1}(t; t_1 \ldots, t_n)} .$$

Note that F_{n+1}, as a regular version of probability distribution, is measurable in $(t, t_1, t_2, \ldots)$, and this yields that φ_{n+1} is measurable as well. The stochastic intensity of the process N is a process $\lambda(t, N)$ such that (see Theorem 2.5.K.),

$$\lambda(t, \mu) = \sum_{n=0}^{\infty} r_{n+1}(t; \tau_1(\mu), \ldots, \tau_n(\mu)) \times \mathbf{1}_{(\tau_n(\mu), \tau_{n+1}(\mu)]}(t) , \tag{1}$$

where $\mu \in \mathcal{N}(R_+)$. That is, $\lambda(t, N)$ is equal to $r_{n+1}(t; \tau_0(N), \ldots, \tau_n(N))$ for $t \in (\tau_n(N), \tau_{n+1}(N)]$.

We shall now recall the construction of the Poisson embedding introduced in Lindvall (1988). Suppose that N is a simple point process with an intensity process $\lambda(t, N)$ defined in (1). Define a mapping $\boldsymbol{\kappa} : N(R_+^2) \to \mathcal{J}_+$ as follows. Let $\boldsymbol{\pi} \in N(R_+^2)$. Set $\kappa_0 = 0$. For $n \geq 1$ let,

$$\begin{aligned} A_{n,t} &= \{(u, v) \in R_+^2 : \kappa_{n-1} < u \leq t,\ v \leq \varphi_n(u; \kappa_1, \ldots, \kappa_{n-1})\} , \\ \kappa_n &= \kappa_n(\boldsymbol{\pi}) = \sup\{t > \kappa_{n-1} : \boldsymbol{\pi}(A_{n,t}) = 0\} . \end{aligned}$$

Now set $\boldsymbol{\kappa}(\boldsymbol{\pi}) = (\kappa_1(\boldsymbol{\pi}), \kappa_2(\boldsymbol{\pi}), \ldots)$. Note that $\boldsymbol{\kappa}$ depends on N via the intensity function $\lambda(t, \mu)$.

Lemma D.

(i) *The mapping* $\boldsymbol{\kappa} : \mathcal{N}(R_+^2) \to \mathcal{J}_+$ *is measurable.*

(ii) *Let* Π *be a Poisson process on* R_+^2 *with the mean measure equal to the Lebesgue measure (the standard Poisson process on* R_+^2*), and set* $N' = \boldsymbol{\tau}^{-1}\boldsymbol{\kappa}(\Pi)$*. Then the point processes* N *and* N' *have the same probability distribution.*

Proof. Part (i) follows from the fact that the mapping $\pi \mapsto \kappa_n(\pi)$ is measurable for each $n \geq 1$. For part (ii) it can be seen that $\boldsymbol{\kappa}(\Pi)$ is equal in distribution to $(T_1, T_2, \ldots)$, hence $N' = \boldsymbol{\tau}^{-1}\boldsymbol{\kappa}(\Pi)$ is equal in distribution to $N = \boldsymbol{\tau}^{-1}(T_1, T_2, \ldots)$. □

Suppose that we have two point processes N_1 and N_2 with intensity processes λ_1, λ_2 and corresponding mappings $\boldsymbol{\kappa}_1$, $\boldsymbol{\kappa}_2$. Suppose also that if $\mu_1 <_{\mathcal{N}} \mu_2$ then $\lambda_1(\tau_n(\mu_2), \mu_1) < \lambda_2(\tau_n(\mu_2), \mu_2)$ for all $n \geq 1$. Then the Poisson embedding $N_1' = \boldsymbol{\tau}^{-1}\boldsymbol{\kappa}_1(\Pi)$ and $N_2' =$

$\boldsymbol{\tau}^{-1}\boldsymbol{\kappa}_2(\Pi)$ yields $N_1' < N_2'$ a.s., and is equivalent to the construction of *thinning* formalized differently e.g., in Rolski and Szekli (1991). Similarly, if the intensity function of say the process N_1 is bounded, i.e., for all t and μ we have $\lambda_1(t,\mu) \leq a$ where $a > 0$, then N_1' is the result of thinning of a Poisson process on R_+ with intensity a.

Let N be a simple point process with an intensity process λ as in (1). We say that N is a **self-exciting** process with respect to the order $<_{\mathcal{N}}$ if and only if for all $\mu_1, \mu_2 \in \mathcal{N}(R_+)$,

$$\mu_1 <_{\mathcal{N}} \mu_2 \Rightarrow \lambda(t,\mu_1) \leq \lambda(t,\mu_2)\ , t > 0. \tag{2}$$

Note that it is sufficient to verify (2) for $t = \tau_n(\mu_2)$, $n \geq 1$.

Theorem E. (Kwiecinski and Szekli (1994)) *If N is self-exciting with respect to the order $<_{\mathcal{N}}$ then N is associated.*

Proof. Suppose that N is a DFR-type process. Then, for all π_1, π_2

$$\pi_1 <_{\mathcal{N}} \pi_2 \Rightarrow \{\kappa_1(\pi_1), \kappa_2(\pi_1), \kappa_3(\pi_1), \ldots\} \subset \{\kappa_1(\pi_2), \kappa_2(\pi_2), \kappa_3(\pi_2), \ldots\}\ ,$$

we omit a proof by induction. We also have, for all $(x_1, x_2, \ldots) \in \mathcal{J}_+$, and $(y_1, y_2, \ldots) \in \mathcal{J}_+$

$$\{x_1, x_2, \ldots\} \subset \{y_1, y_2, \ldots\} \Rightarrow \boldsymbol{\tau}^{-1}(x_1, x_2, \ldots) < \boldsymbol{\tau}^{-1}(y_1, y_2, \ldots)\ .$$

It follows easily that $\boldsymbol{\tau}^{-1}\boldsymbol{\kappa}$ is increasing, that is, for all π_1, π_2

$$\pi_1 <_{\mathcal{N}} \pi_2 \Rightarrow \boldsymbol{\tau}^{-1}\boldsymbol{\kappa}(\pi_1) < \boldsymbol{\tau}^{-1}\boldsymbol{\kappa}(\pi_2)\ .$$

Let Π be the standard Poisson process in R_+^2, and let $N' = \boldsymbol{\tau}^{-1}\boldsymbol{\kappa}(\Pi)$. Since Π is associated and $\boldsymbol{\tau}^{-1}\boldsymbol{\kappa}$ is increasing then N' is also associated (Theorem 3.2, Lindqvist (1988)). □

Since clearly a renewal process with DFR renewal distribution is self-exciting with respect to $<_{\mathcal{N}}$, Theorem E. yields the following result, which generalizes Theorem 1 of Burton and Waymire (1986)

Corollary F. *The stationary distribution and the Palm distribution of a renewal process N on R_+ with a continuous DFR lifetime distribution are associated.*

From 2.11. (4) it is immediate that the failure point process for the block replacement policy $N^{B,T}$ with DFR life times is self-exciting, hence we have

Corollary G. *The failure point process of block replacement policy $N^{B,T}$ with $F \in DFR$ is associated*

Appendix A

A.1 Probability spaces

In probability theory the starting point is a "probability space". A usual phrase at the beginning of a stochastic model is (or should be) :" Let $(\Omega, \mathcal{F}, \mathbf{P})$ be a probability space...". Such a general approach to **probability** plays a fundamental role in the theory, and it is not our intention to recall all definitions and axioms, which can be easily found in any textbook on probability theory. We shall however recall a number of examples of probability spaces which are of importance for stochastic ordering. We shall see that nothing is really random in these examples of probability spaces; the specific flavor of probability theory appears later with the concept of **independence.**

Example(a). The structurally simplest example is

$$\Omega = \{1, \ldots, M\},\ M \in \mathbf{N},\ \mathcal{F} = 2^{\Omega},\ \mathbf{P},$$

where 2^{Ω} denotes the set of all subsets of Ω, and $\mathbf{P}$ is any function defined on $\mathcal{F}$ with values in the interval [0,1], which is additive i.e. $\mathbf{P}(A \cup B) = \mathbf{P}(A) + \mathbf{P}(B)$, when A and B are disjoint subsets of Ω.

This probability space serves as a model for a number of **experiments**, where the set of outcomes is finite. In this case, for example, Ω represents possible **elementary outcomes** or **events**, $\mathcal{F}$ is a collection of all possible events, and $\mathbf{P}$ is a measure of "proportion" of the number of **elementary events** which describe a given event A $(A \subseteq \Omega)$, to the number of all possible elementary events. This class of models includes **experiments** describing coin tossing, playing cards, throwing dice, placing balls etc. The usual difficulty in this model consists of the inevitable use of combinatorics.
Formally, the situation is simple; each $\mathbf{P}$ is described by a sequence of [0,1] -valued numbers : $p_1, \ldots, p_M$, such that $p_1 + \ldots + p_M = 1$. In particular situations of interest we obtain numbers which provide a probabilistic description of the model. The variety of particular interesting cases is very reach (see e.g. Feller (1968)).

A probability space requires that

- $\Omega \in \mathcal{F}$,
- $A \in \mathcal{F} \Rightarrow A^c \in \mathcal{F}$ $(A^c = \Omega \setminus A)$,
- $A_i \in \mathcal{F}, i = 1, 2, \ldots \Rightarrow \bigcup_{i=1}^{\infty} \in \mathcal{F}$.

A family of sets $\mathcal{F}$ fulfilling the above conditions is called a σ–field. The smallest σ–field containing a given family of sets $\mathcal{A}$ is called the σ–field generated by $\mathcal{A}$, which is denoted by $\sigma(\mathcal{A})$. It is the intersection of all σ–fields containing $\mathcal{A}$. If Ω is equal to $[0,1]$ then the σ–field generated by all open intervals of $[0,1]$ is called the Borel σ–field of $[0,1]$.

Example(b). $\Omega = [0,1], \mathcal{F} = \mathcal{B}_{[0,1]}, \mathbf{P} = \ell,$

where $\mathcal{B}_{[0,1]}$ denotes the Borel σ-field on $[0,1]$, ℓ- denotes the Lebesgue measure on [0,1].

This example is of importance in constructing mathematical models, especially when in use are random variables uniformly distributed on [0,1]. At the same time it may be the first major obstacle of theoretical nature, because it requires a use of σ-fields of sets and measures defined on σ-fields. The notion of a σ-field of sets is simple and it does not cost a lot of mental energy to accept it, however extending a measure on a σ-field requires some technical work (**the measure extention theorem**).
On the other hand this example is in a sense natural, because roughly speaking the Borel σ-field consists of all intervals and sets which can be obtained from open intervals by consecutive applications of countable sums and intersections. This is a reasonable collection of sets about which we can say that they can be measured. The Lebesgue measure ℓ in turn is a measure which corresponds to our intuitions of measuring a set by its length (the Lebesgue measure of an interval is equal to its length). Of course for more complicated sets than intervals we need a more delicate approach. We can not measure all possible subsets of [0,1], hence we define a collection of appropriate sets related to intervals, and we introduce the concept of σ-fields of sets.

Example(c). $\Omega = R_+, \mathcal{F} = \mathcal{B}^1_+, \mathbf{P} = \mathbf{P}_{\mathbf{exp}},$

where R_+denotes nonnegative real numbers, $\mathcal{B}^+_1$ is the Borel σ-field on R_+, $\mathbf{P}_{\mathbf{exp}}$ denotes the probability measure, which on intervals is defined by:

$$\mathbf{P}_{\mathbf{exp}}((a,b)) = e^{-a} - e^{-b},$$

for $a, b \in R_+$.

The measurable structure of this example is similar to that of Example(b), i.e. the Borel σ-field is taken as the σ-field of events. However now infinite intervals have positive measure. This example will be useful to represent exponentially distributed lifetimes in some systems.

The way the measure is introduced here is a special case of the general situation, where

$$\mathbf{P}((a,b)) = F(b) - F(a),$$

for some nondecreasing real function F, with $\lim_{x\to\infty} F(x) = 1, \lim_{x\to-\infty} F(x) = 0$.
This is a basic relation which allows us to consider real functions defined on the real axis instead of probability measures on σ-fields. This relationship will be considered in more detail in the next section.

Each of the previous examples can be extended by taking so called **product spaces.**

Example($\bar{a}$). $\Omega = \{1, \ldots, M\}^n, \mathcal{F} = 2^\Omega, \mathbf{P}^{(n)} = \mathbf{P} \times \ldots \times \mathbf{P}$ (n times), where $\{1,\ldots,M\}^n$ denotes the product of n sets $\{1,\ldots,M\}$. Each element of $\{1,\ldots,M\}^n$ is a finite sequence of a length n , with values in $\{1,\ldots,M\}$. The measure $\mathbf{P}^{(n)}$ is the product of n measures $\mathbf{P}$, i.e.

$$\mathbf{P}^{(n)}(A_1 \times \ldots \times A_n) = \mathbf{P}(A_1)\cdots\mathbf{P}(A_n),$$

for all $A_1,\ldots,A_n \subseteq \{1,\ldots,M\}$, where $\mathbf{P}$ is as given in Example(a).

Example($\bar{b}$). $\Omega = [0,1]^n, \mathcal{F} = \mathcal{B}^n_{[0,1]}, \mathbf{P} = \ell_n,$

where $\mathcal{B}^n_{[0,1]}$ denotes the product σ-field of n Borel σ-fields $\mathcal{B}_{[0,1]}$ from Example(b), i.e. the smallest σ-field containing product sets of the form $A_1 \times \ldots \times A_n$; $A_1,\ldots,A_n \in \mathcal{B}_{[0,1]}$. The measure ℓ_n is the n dimensional Lebesgue measure.

Example($\bar{c}$). $\Omega = R^n_+,\ \mathcal{F} = \mathcal{B}_n,\ \mathbf{P}^{(n)}_{\exp} = \mathbf{P}_{\exp} \times \ldots \times \mathbf{P}_{\exp}$ (n times),

where $\mathcal{B}_n$ denotes the n dimensional Borel σ-field, which is the product of n one dimensional Borel σ-fields $\mathcal{B}_1$ from Example(c), and $\mathbf{P}^{(n)}_{\exp}$ is the product measure of n measures $\mathbf{P}_{\exp}$ from Example(c), i.e.

$$\mathbf{P}^{(n)}_{\exp}(A_1 \times \ldots \times A_n) = \mathbf{P}_{\exp}(A_1)\ldots\mathbf{P}_{\exp}(A_n),$$

for $A_1,\ldots,A_n \in \mathcal{B}_1$,. In particular

$$\mathbf{P}^{(n)}_{\exp}((a_1,b_1) \times \ldots \times (a_n,b_n)) = (e^{-a_1} - e^{-b_1})\ldots(e^{-a_n} - e^{-b_n}),$$

for all $a_i, b_i \in R_+, i = 1,\ldots,n.$

A.2 Distribution functions

We begin by recalling the definition of a distribution function on R.

Definition A. *The distribution function is a function $F : R \to R$ which is nondecreasing, right-continuous with* $\lim_{x\to\infty} F(x) = 1$, *and* $\lim_{x\to-\infty} F(x) = 0$.

Remark. There is 1-1 correspondence between the set of probability measures on $(R, \mathcal{B}^1)$ and the set of distribution functions on R.

We do not prove this result. Instead, we check how these notions are related to Examples(a)-(c).

Example(a'). Consider the probability space $(\{1,\ldots,M\}, 2^{\{1,\ldots,M\}}, \mathbf{P})$, where $\mathbf{P}(\{i\}) = p_i, i = 1,\ldots,M$ for some collection of numbers $\{p_i\}_{i=1,\ldots,M}$, with $\sum_{i=1}^M p_i = 1, 0 \le p_i \le 1, i = 1,\ldots$,M.
The probability measure $\mathbf{P}$ is completely determined by the sequence $\{p_i\}$. With this measure we associate the following distribution function

$$F(t) = \mathbf{P}(\{1,\ldots,M\} \cap (-\infty, t]),\ t \in R.$$

Thus,

$$F(t) = \sum_{i \le t} p_i$$

which implies that F is a distribution function. It is clear that a distribution function of the above form defines a probability measure on $\{1,\ldots,M\}$ by $\mathbf{P}(\{i\}) = p_i$. Thus, there exists one-to-one correspondence between the class of probability measures on $\{1,\ldots,M\}$ and the class of distribution functions of the above form. This example can be easily generalized if we take an arbitrary collection of real values $\{x_1, x_2, \ldots : x_1 < x_2 < \ldots\}$ instead of $\{1,\ldots,M\}$. The resulting distribution functions are called **discrete distribution functions.**

Example(b'). Consider now the probability space $([0,1], \mathcal{B}_{[0,1]}, \ell)$. The corresponding distribution function we define analogously to Example(a') by

$$F(t) = \ell([0,1] \cap (-\infty, t]), t \in R. \tag{1}$$

Thus we have

$$F(t) = \begin{cases} 1 & t > 1, \\ t & 0 < t \leq 1, \\ 0 & t \leq 0. \end{cases}$$

Obviously, F is a distribution function. The corresponding measure gives the value t for each interval $[0,t]$, when $t \in [0,1]$, i.e., its length. There exists only one Borel measure with this property: the Lebesgue measure. The above distribution function is called the **uniform** distribution on $[0,1]$. The corresponding probability measure to the uniform distribution function on $[0,1]$ is the Lebesgue measure on $[0,1]$. This example can be generalized to arbitrary intervals $[a,b]$, $a, b \in R$, and the **uniform distribution function** on $[a,b]$.

Example(c'). For $(R_+, \mathcal{B}^1_+, \mathbf{P}_{\mathbf{exp}})$ we have

$$F(t) = \mathbf{P}_{\mathbf{exp}}(R_+ \cap (-\infty, t]) = \begin{cases} 1 - e^{-t} & t > 0 \\ 0 & t \leq 0 \end{cases}$$

This distribution is called the standard **exponential** distribution function. It corresponds to the measure $\mathbf{P}_{\mathbf{exp}}$. This example can be generalized to arbitrary exponential distribution functions of the form

$$F(t) = 1 - e^{-\lambda t}, \ t > 0,$$

where $\lambda > 0$ is a scale parameter.

Multidimensional distribution functions are natural generalization of one dimensional distribution functions. Indeed in the one dimensional case we see from Examples $(a') - (c')$ that $F(t)$ is the value of the measure of $\Omega \cap (-\infty, t]$. The values on arbitrary intervals $\Omega \cap (s, t]$ are given by $F(t) - F(s)$. For the probability spaces from the examples $(\bar{a}) - (\bar{c})$, i.e. for multidimensional sets, probability measures are completely determined by the respective values on multidimensional intervals. Suppose that Ω is a subset of R^n , and $\mathbf{P}$ is a probability measure on Ω. Consider a multidimensional interval $(-\infty, \mathbf{t}] = (-\infty, t_1] \times \ldots \times (-\infty, t_n]$, where $\mathbf{t} = (t_1, \ldots, t_n), t_i \in R, i = 1,\ldots$,n. The value $\mathbf{P}((-\infty, \mathbf{t}])$ determines a function $\mathbf{F}(\mathbf{t})$ with some special properties. This properties we use as the defining properties of multidimensional distribution functions. Let us start with the 2-dimensional case.

Definition B. *The 2-dimensional distribution function is a function* $\mathbf{F} : R^2 \to R$*, for which*
(i) $\mathbf{F}$ *is nondecreasing in each variable,*
(ii) $\mathbf{F}$ *is right continuous,*
(iii)

$$\lim_{y\to\infty}\lim_{x\to\infty}\mathbf{F}(x,y) = \lim_{x\to\infty}\lim_{y\to\infty}\mathbf{F}(x,y) = 1,$$

$$\lim_{x\to-\infty}\lim_{y\to-\infty}\mathbf{F}(x,y) = \lim_{y\to-\infty}\lim_{x\to-\infty}\mathbf{F}(x,y) = 0, \ x,y \in R,$$

(iv)

$$\mathbf{F}(x,y) + \mathbf{F}(x',y') \geq \mathbf{F}(x,y') + \mathbf{F}(x',y),$$

where $x \leq x'$, $y \leq y'$.

The condition (iv) means that the corresponding probability measure defined by $\mathbf{P}((-\infty,\mathbf{t}]) = \mathbf{F}(\mathbf{t}), \mathbf{t} \in R^2$ should have nonnegative value on the 2-dimensional interval $(x,x'] \times (y,y']$.
Functions which fulfill the condition (iv) are called **superadditive**.

In an arbitrary finite dimension, the formalism is most conveniently described by n-positive functions which are defined as follows

Definition C. *A function* $\phi : R^n \to R$ *is* n*-positive if*

$$\Delta^n_{x_n,x'_n}\cdots\Delta^1_{x_1,x'_1}\phi \geq 0,$$

for all $\mathbf{x} \leq \mathbf{x}'$*, where*

$$\Delta^k_{x_k,x'_k}\phi = \Phi(x_1,\ldots,x_{k-1},x'_k,x_{k+1},\ldots,x_n) - \Phi(x_1,\ldots,x_{k-1},x_k,x_{k+1},\ldots,x_n),$$

$\mathbf{x} = (x_1,\ldots,x_n)$.

1-positive functions are nondecreasing.
2-positive functions are superadditive.

Now we define general multivariate distribution functions.

Definition D. *The* n*-dimensional distribution function is a function* $\mathbf{F} : R^n \to R$*, which is right continuous, with*
(c) $\lim_{\mathbf{x}\to\infty}\mathbf{F}(\mathbf{x}) = 1$, (c) $\lim_{\mathbf{x}\to-\infty}\mathbf{F}(\mathbf{x}) = 0, i = 1,\ldots,n,\ \mathbf{x} \in R^n$,
and which is n*-positive*
(here (c) $\lim_{\mathbf{x}\to\infty}$ *stands for* $\lim_{x_{i_1}\to\infty}\cdots\lim_{x_{i_n}\to\infty}$*, for an arbitrary permutation* $(i_1,\ldots,i_n)$ *of* $(1,\ldots,n)$*).*

Example($\bar{a}'$). $\Omega = \{1,\ldots,M\}^n$, $\mathcal{F} = 2^\Omega$, $\mathbf{P}^{(n)} = \mathbf{P}\times\ldots\times\mathbf{P}$,
here

$$\mathbf{F}(\mathbf{t}) = \mathbf{P}^{(n)}(\Omega \cap (-\infty,\mathbf{t}]).$$

From examples (a) and ($\bar{a}$) we see that

$$\mathbf{F}(\mathbf{t}) = \sum_{p_{i_k}\leq t_k} p_{i_1}\cdots p_{i_n},$$

where $\{p_i\}_{i\geq 1}$ describes the probability measure $\mathbf{P}$ on $\{1,\ldots,M\}^n$.
Alternatively, we can write

$$\mathbf{F}(\mathbf{t}) = F(t_1)\ldots F(t_n),$$

where F is from Example (a').

Example($\bar{b}'$). $\Omega = [0,1]^n, \quad \mathcal{F} = \mathcal{B}^n_{[0,1]}, \quad \mathbf{P} = \ell_n.$

$$\mathbf{F}(\mathbf{t}) = \ell_n([0,1] \cap (-\infty, \mathbf{t}]), \qquad \mathbf{t} \in R^n,$$

or alternatively

$$\mathbf{F}(\mathbf{t}) = F(t_1)\cdots F(t_n),$$

for F from Example (b').

Example($\bar{c}'$). $\Omega = R^n_+, \quad \mathcal{F} = \mathcal{B}^n, \quad \mathbf{P}^{(n)}_{\mathbf{exp}} = \mathbf{P}_{\mathbf{exp}} \times \ldots \times \mathbf{P}_{\mathbf{exp}}$ (n times).
Here the corresponding multidimensional distribution, as in the previous example is

$$\mathbf{F}(\mathbf{t}) = F(t_1)\cdots F(t_n),$$

where F is from Example (c'), i.e.

$$\mathbf{F}(\mathbf{t}) = (1 - e^{-t_1})\cdots(1 - e^{-t_n}),$$

for $t_1, \ldots, t_n \geq 0$, and 0 otherwise.

A.3 Examples of distribution functions

The following distribution functions are defined for $t \in$R.

Bernoulli:

$$F(t) = (1-p)\delta_0(t) + p\delta_1(t),$$

where $p \in (0,1), t \in R$, and

$$\delta_x(t) = \begin{cases} 0 & t < x, \\ 1 & t \geq x \end{cases}$$

for $x \in R$.

Binomial:

$$F(t) = \sum_{k=0}^{n} \binom{n}{k} p^k (1-p)^{n-k} \delta_k(t),$$

where $p \in (0,1), n \in$N.
The expected value of this distribution equals np, variance $np(1-p)$. The binomial distribution is the n fold convolution of Bernoulli distributions, that is the distribution of the number of successes in n Bernoulli trials.

Geometric:

$$F(t) = \sum_{k=0}^{\infty} p(1-p)^k \delta_k(t),$$

where $p \in (0,1)$.

See Section 1.7 for more details.

Poisson:

$$F(t) = \sum_{k=0}^{\infty} (\lambda^k e^{-\lambda}/k!)\delta_k(t),$$

where $\lambda > 0$.
This is the distribution of the number of events in the $[0, 1]$ interval in homogeneous Poisson process with λ intensity. This is also a limiting distribution for binomial distributions if $np_n \to \lambda,\ n \to \infty$.

Negative binomial:

$$F(t) = \sum_{k=0}^{\infty} \binom{-\alpha}{k} p^{\alpha}(p-1)^k \delta_k(t),$$

where $\alpha > 0, p \in (0, 1)$, and

$$\binom{-\alpha}{k} = \frac{-\alpha(-\alpha-1)\ldots(-\alpha-k+1)}{k!}$$

The number of successes in Bernoulli trials before a fixed number of failures has a negative binomial distribution.

Erlang:

$$F(t) = \delta_0(t)[1 - e^{-\lambda t}(1 + \ldots + (\lambda t)^{n-1}/(n-1)!)]$$

where $\lambda > 0, n \in \mathrm{N}$.

Gamma:

$$F(t) = \delta_0(t)(\int_0^t \lambda(\lambda x)^{\beta-1} e^{-\lambda x} dx)/\Gamma(\beta),$$

where $\lambda > 0, \beta > 0$, and

$$\Gamma(\beta) = \int_0^{\infty} \lambda(\lambda x)^{\beta-1} e^{-\lambda x} \mathrm{dx}.$$

Normal:

$$F(t) = \int_{-\infty}^{t} e^{-(x-m)^2/2\sigma^2}/\sigma(2\pi)^{1/2} dx,$$

where $m \in R, \sigma > 0$.

Weibull:

$$F(t) = \delta_0(t) \int_0^t \lambda \alpha x^{\alpha-1} e^{-\lambda x^{\alpha}} dx,$$

where $\lambda > 0, \alpha > 0$.

A.4 Other characteristics of probability measures

We usually use distribution functions to describe probability measures. However on R, probability measures are often well described by densities, failure rates, Laplace transforms, characteristic functions, or generating functions. Of course all of them are related to a given distribution function, and each of them determines uniquely the corresponding probability measure.
To introduce a complete theory of densities it would be necessary to introduce integration with respect to general measures on Borel σ-fields (see e.g. Billingsley (1986)). To avoid such an extensive description and to use only a Riemann or a Riemann-Stieltjes integration, we consider only particular cases.

Densities

In the case when a probability measure $\mathbf{P}$ is described by a discrete distribution function (see Example (a')), i.e. it is concentrated on a collection of reals $x_1 < \ldots < x_M$, the corresponding probability measure is defined by a sequence $\{p_i\}_{i=1,\ldots,M}$, such that, $0 \leq p_i \leq 1$, $\sum_{i=1}^{M} p_i = 1$ (M can be infinite), where $p_i = \mathbf{P}(\{x_i\})$. The function $p : N \to [0,1]$, $p(i) = p_i$ is called **discrete density function, counting density function**, or **probability mass function**. Recall that it is related to the corresponding distribution function F by

$$F(t) = \sum_{x_i \leq t} p_i,$$

$$= \sum_{i=1}^{M} p_i \delta_{x_i}(t) = \sum_{i=1}^{M} p_i \delta_{x_i}(t).$$

Note that the Bernoulli, binomial, geometric, Poisson, negative binomial distribution functions are of this type.

If a cumulative distribution function F has the form

$$F(t) = \int_{-\infty}^{t} f(x)dx,$$

then f is called the **density function** of a continuous distribution function. Here the integration can be understood in the Riemann sense. Note that normal, gamma, Weibull, exponential, uniform, Erlang distribution functions are of this type.
The function f has a graphic interpretation; if we plot the function f in the 2-dimensional Cartesian coordinates (x, y), the area bounded by the plot of f, the axis OX and two vertical lines crossing OX at a, b ($a < b$), determines the value of the corresponding probability measure $\mathbf{P}$ on (a, b), i.e. $\mathbf{P}((a, b))$.

Failure rates

Consider a discrete distribution function F , with a density $\{p_i\}$. The failure rate function is

$$r(k) = p_k / \sum_{j=k}^{\infty} p_j,$$

provided $\sum_{j=k}^{\infty} p_j \geq 0$.

The negative binomial distributions have increasing failure rates for $\alpha > 1$, and decreasing failure rates for $\alpha < 1$. This family coincides with the geometric family for $\alpha = 1$, and, in this case, the failure rate is constant. The binomial and Poisson distributions have increasing failure rates.

If a continuous distribution function F, with $F(0) = 0$ (we call such distributions **life distributions**), has density f, then the failure rate function $r(t)$ is defined for the values of t, for which $F(t) < 1$, by

$$r(t) = f(t)/\bar{F}(t),$$

where $\bar{F}(t) = 1 - F(t)$.
This function has a useful probabilistic interpretation; $r(t)dt$ represents the probability that an object of age t will fail in the interval $[t, t + dt]$. It is important in a number of applications and known by a variety of names. For example "force of mortality" in actuarial sciences; "intensity function" in extreme value theory; "hazard rate" in reliability. For an extensive treatment of failure rates see Barlow and Proschan (1981). From the definition of the failure rate we have

$$\bar{F}(t) = \exp\{-\int_0^t r(x)dx\},$$

for t such that $F(t) < 1$. The function $R(t) = \int_0^t r(x)dx$ is called the **total failure rate.**

Transforms

Using the concept of integration with respect to a probability measure or with respect to a distribution function we obtain a powerful description of probability measures on R. This integration is applied to some special families of functions. For example taking for such a family $\{f_\lambda(x) = e^{-\lambda x}, \lambda > 0\}$, we define the Laplace transform, which is $\phi(\lambda) = \int_0^\infty e^{-\lambda x} dF(x)$, for distribution functions concentrated on $(0, \infty)$. Using different classes of functions and the methods of integration, allows us to introduce different transformations of probability measures. The proof that such transformations uniquely determine the corresponding probability measure is based on the Stone-Weierstrass theorem. Because we will use such transformations only occasionally we refer the interested reader to Widder(196?) and Lukacs (1960), for a more detailed description of the theory.

Among discrete distribution functions those concentrated only on non-negative integer values are of special importance. Their study is facilitated by the powerful method of generating functions which is rarely fully utilized.

Definition A. *Let $\{p_i\}_{i \geq 0}$ be a discrete density function.*

$$P(s) = p_0 + p_1 s + p_2 s^2 + \dots,$$

which is convergent absolutely at least for $-1 \leq s \leq 1$, is called **generating function** *of $\{p_i\}$.*

The continuous analog of generating functions is the concept of Laplace transforms. Suppose that F is a life distribution with a density f. The **Laplace transform** of F is the function defined by

$$\phi_F(\lambda) = \int_0^\infty e^{-\lambda x} f(x)dx, \ \lambda > 0.$$

A.5 Random variables equal in distribution

We have recalled basic characteristics describing probability measures on R or R^n. The special interest on probability measures on these spaces is caused by the fact that in descriptions of applied probability models we usually use real valued characteristics or finite collections of them, called random variables.

Definition A. *Suppose* $(\Omega, \mathcal{F}, \mathbf{P})$ *is a probability space. A function* $X : \Omega \to R$ *such that* $\{\omega : X(\omega) \leq a\} \in \mathcal{F}, a \in R, \omega \in \Omega$, *is a random variable.*

Corollary B. *For the discrete probability space from the Example (a) any function* $X : \{1, \ldots, M\} \to R$ *is a random variable.*

Because we assume that for a random variable X , the set $\{\omega : X(\omega) \leq a\} \in \mathcal{F}$, this set can be measured, i.e. $\mathbf{P}(\{\omega : X(\omega) \leq a\})$ exists (we will write shorter $\mathbf{P}(X \leq a)$). Thus we have the following lemma.

Lemma C. *The function*

$$F_X(t) = \mathbf{P}(X \leq t), \ t \in R,$$

is a distribution function on R.

We call $F_X(t)$ the cumulative distribution function of the random variable X.
Thus each random variable has a corresponding probability measure $\mathbf{P}_X$ on R (or on a subset of R). Such a probability measure is called the distribution of X.

For finite collections of random variables we consider distributions on R^n. Let $\mathbf{X} = (X_1, \ldots, X_n), \mathbf{t} = (t_1, \ldots, t_n)$, and $\mathbf{x} \leq \mathbf{t} \Leftrightarrow x_i \leq t_i, i = 1, \ldots$,n.

Lemma D. *The function*

$$\mathbf{F_X(t)} = \mathbf{P(X \leq t)}, \mathbf{t} \in R^n,$$

is a multidimensional distribution function on R^n.

We call $\mathbf{F_X}$ the distribution function of the random vector $\mathbf{X}$.

Example. $\Omega = [0,1], \mathcal{F} = \mathcal{B}_{[0,1]}, \mathbf{P} = \ell$.
Let $X_1(\omega) = \omega, \quad X_2(\omega) = 1 - \omega, \quad \omega \in \Omega$.
Of course X_1 and X_2 are random variables on Ω. We can easily find their distribution functions. Let $t \in [0,1]$.

$$\mathbf{P}(X_1 \leq t) = \ell(\{\omega : \omega \leq t\}) = \ell([0,t]) = t$$

$$\mathbf{P}(X_2 \leq t) = \ell(\{1 - \omega \leq t\}) = \ell([1-t, 1]) = \mathrm{t}.$$

Hence the corresponding distribution functions F_{X_1}, F_{X_2} are equal, both of them being uniform on [0,1].

Two different random variables can have the same distribution.

Let $(\Omega_i, \mathcal{F}_i, \mathbf{P}_i), i = 1,2$ be probability spaces, and $\mathbf{X}_i$, random variables with values in R^n defined on these spaces. (We adopt the name "random variables" for random vectors and for random elements with values in general spaces).

Definition E. *Random variables* $\mathbf{X}_1$ *and* $\mathbf{X}_2$ *are are equal in distribution if*
$\mathbf{P}_1(\mathbf{X}_1 \leq \mathbf{a}) = \mathbf{P}_2(\mathbf{X}_2 \leq \mathbf{a})$, *for* $\mathbf{a} \in R^n$.

If two random variables are equal in distribution then we say that one is a **version** of the other.

A.6 BIBLIOGRAPHY

Ahmed, N., Leon, R. and Proschan, F. (1981) Generalized association with applications in multivariate statistics. Ann.Stat.9,168-176.

Alfsen, E. (1971) *Compact Convex Sets and Boundary Integrals*. Springer, New York.

Arjas,E. and Lehtonen, T. (1978) Approximating many server queues by means of single server queues. Math. Operat. Res. 3, 205-223.

Arjas, E. and Norros, I. (1984) Life lengths and association: a dynamic approach. Math. Operat. Res. 9, 151-158.

Asmussen, S.(1987) *Applied Probability and Queues*, John Wiley & Sons, Chichester.

Baccelli, F. and Bremaud,P. (1993) *Elements of Queueing Theory*. Springer Verlag.

Bartoszewicz, J. (1986) Dispersive ordering and the total time on test transformation. Stat. Prob. Letters 4, 285-288.

Baskett,F., Chandy,K.M., Muntz,P.R. and Palacios,F.G.(1975) Open, closed and mixed networks of queues with different classes of customers. J.Assoc. Comp. Mach. 22, 248-260.

Barlow, R.E. and Proschan, F. (1981) *Statistical Theory of Reliability and Life Testing: Probability Models*, To Begin With, Silver Spring, Maryland.

Beichelt, F. (1993) A unifying treatment of replacement policies with minimal repair. Naval Research Logistics 40, 51-67.

Benes, V.E. (1959) On trunks with negative exponential holding times serving a renewal process. Bell System Tech. J. 38, 211-258.

Billingsley, P. (1986) *Probability and Measure*, 2nd Edition, John Wiley & Sons, New York.

Blackwell, D. (1953) Equivalent comparisons of experiments. Ann. Math. Statist. 24, 265-272.

Block, H.W., Langberg, N. and Savits, T.H. (1990) Maintenance comparisons: Block policies. J. Appl. Prob. 27, 649-657.

Block, H.W., Langberg, N. and Savits, T.H. (1993) Repair replacement policies. J. Appl. Prob. 30, 194-206.

Block, H.W., Savits, T.H. and Shaked, M. (1982) Some concepts of negative dependence. Ann.Prob.10,765-772.

Borovkov, A.A. (1976) *Stochastic Processes in Queueing Theory*, Springer-Verlag, New York.

Bremaud, P. (1981) *Point Processes and Queues: Martingale Dynamics*. Springer, New York.

Brown, M. (1980) Bounds, inequalities and monotonicity properties for some specialized processes. Ann. Prob. 8, 227-240.

Brown, M. (1990) Error bounds for exponential approximations of geometric convolutions. Ann. Prob. 18, 1388-1402.

Brown, T.C. and Nair, M.G. (1988) A simple proof of the multivariate random time change theorem for point processes. J. Appl. Prob. 25, 210-214.

Bruckner, A.M. and Ostrow, E. (1962) Some function classes related to the class of convex functions. Pacific J. Math. 12, 1203-1215.

Bruijn, N.G. de and Erdős, P. (1953) On a recursion formula and some Tauberian theorems. J. Research Nat. Bureau Stand. 50, 161-164.

Burke, P.J. (1956) The output of a queuing system. Oper. Res. 4, 699-704.

Burton, R.M. and Waymire, E. (1986) A sufficient condition for association of a renewal process. Ann. Prob. 14, 1272-1276

Chang, C.S. (1989) Comparison theroems for queueing systems and their applications to ISDN. PhD Dissertation, Electr. Eng. Columbia Univ. New York.

Chang, C.S. and Pinedo, M. (1990) Bounds and inequalities for single server loss systems. Queueing Systems 6, 425-436.

Chang, C.S. (1992) A new ordering for stochastic majorization: theory and applications. Adv. Appl. Prob. 24, 604-634.

Chung, K.L. (1967) *Markov Chains with Stationary Transition Probabilities*, 2nd Edition, Springer-Verlag, New York.

Chung, K.L. (1974) *A Course in Probability Theory*, 2nd ed. Academic Press, New York.

Chung, K.L. (1982) *Lectures from Markov Processes to Brownian Motion*, Springer-Verlag, New York.

Cohen, J.W. (1969) *The Single Server Queue*, North Holland, Amsterdam.

Cooper, R. and Palakurthi, S. (1989) Heterogeneous server loss systems with ordered entry:an anomaly. Operat. Res. Letters 8, 347-349.

Cox, D.R. and Isham, V. (1980) *Point Processes*, Chapman and Hall, London.

Daduna, H. and Szekli, R. (1992) Dependencies in Markovian networks. Adv. Appl. Prob..

Daduna, H. and Szekli, R. (1992) A queueing theoretical proof of an increasing property of Polya frequency functions. Preprint.

Daley, D.J. (1977) Inequalities for moments of tails of random variables with a queueing application. Zeitsch. Wahrsch. Geb. 41, 139-143.

Daley, D.J. and Moran, P.A.P. (1968) Two sided inequalities for waiting time and queue size in GI/G/1. Theory Prob. Appl. 13, 356-358.

Daley, D.J. and Vere-Jones, D. (1988) *An Introduction to the Theory of Point Processes*, Springer, New York.

Dharmadhikari, S.W. and Joag-dev, K. (1988) *Unimodality, Convexity, and Applications*, Academic Press, Boston.

Disney, R. , Kiessler,P.C.(1987) *Traffic Processes in Queueing Networks, A Markov Renewal Approach.* The Johns Hopkins University Press, London.

Doob, J.L. (1953) *Stochastic Processes*, Wiley, New York.

Dynkin, E.B. (1965) *Markov Processes*, Vol.I., Springer-Verlag, Berlin.

Efron, B. (1965) Increasing properties of Polya frequency functions. Ann.Stat. 272-279.

Elton, J. and Hill, T.P. (1992) Fusions of a probability distribution. Ann. Prob. 20, 421-454.

Erlang, A. (1917) Solution of some problems in the theory of probabilities of significance in automatic telephone exchanges. Post Office Electrical Engineer's Journal 10, 189-197.

Esary, J.D., Marshall, A.W. and Proschan F. (1973) Shock models and wear processes. Ann. Prob. 1, 627-649.

Esary,J.D., Proschan,F., Walkup,D.(1967) Association of random variables with applications. Ann.Math.Stat.38,1466-1474.

Ethier, S.N. and Kurtz, T.G. (1986) *Markov Processes, Characterization and Convergence*, John Wiley & Sons, New York.

Feller, W. (1940) On the integro-differential equations of purely discontinuous Markoff processes. Trans. Amer. Math. Soc. 48, 488-515.

Feller, W. (1968) *An Introduction to Probability Theory and Its Applications*, Vol. 1. 3rd ed. Wiley, New York.

Feller, W. (1971) *An Introduction to Probability Theory and Its Applications* Vol.2, 2nd ed. Wiley, New York.

Fleischmann, K. (1976) Optimal input for the loss system G/M/2. Math. Operationsforsch. Statist. 1, 129-137.

Foley, R.D., Kiessler, P.C. (1989) Positive correlations in a three node Jackson queueing network . Adv.Appl.Prob.21,241-242.

Fortuin, C., Kastelyn, P. and Ginibre, J. (1971) Correlation inequalities on some partially ordered sets. Comm. Math. Phys. 22, 89-103.

Franken, P., Kőnig, D., Arndt, U. and Schmidt, V. (1981) *Queues and Point Processes*, Akademie-Verlag, Berlin.

Gaede, K.W. (1965) Konfidenzgrenzen bei Warteschlangen-und-lagerhaltungs problem. Zeitsch. Angew. Math. Mech. 45, T91-T92.

Gertsbakh, I.B. (1984) Asymptotic methods in reliability theory: A review. Adv. Appl. Prob. 16, 147-175.

Glasserman, P. and Gong, W. (1991) Time changing and truncating K-capacity queues from one to another. J. Appl. Prob. 28, 647-655.

Gordon, W.J. , Newell, G.F. (1967) Closed queueing systems with exponential server. Oper. Res. 15 ,254-265.

Grandell, J. (1976) *Doubly Stochastic Poisson Processes*, Springer Lecture Notes Math. 529, Springer-Verlag, New York.

Guy, D.L. (1961) Common extensions of finitely additive probability measures. Porth. Math. 20, 1-5.

Hansen, B.G. (1990) Some monotonicity properties of the compound geometric distribution and of the renewal function. Manuscript.

Hardy, G.H., Littlewood, J.E. and Polya, G. (1952) *Inequalities.* 2nd ed. Cambridge Univ. Press, London/New York.

Harris,T.G.(1977) A correlation inequality for Markov processes in partially ordered spaces. Ann.Prob.5, 451-454.

Harris, C.M. and Prabhu, N.U. (1987) Stochastic comparisons for single server queues. Naval Res. Logistics 34, 555-567.

Jackson, J.R. (1957) Networks of waiting lines. Oper. Res. 5, 518-521.

Jacobs, P.A. (1986) First passage times for combinations of random loads. SIAM J. Appl. Math. 46, 643-656.

Jacobs, D.R. and Schach, S. (1972) Stochastic order relationships between GI/G/k systems. Annals of Math. Stat. 43, 1622-1633.

Jacod, J. (1975) Multivariate point processes: Predictable projection, Radon-Nikodym derivatives, Representation of martingales. Z. Wahrsch. Verw. Geb. 34, 235-253.

Jean-Marie, A. and Liu, Z. (1992) Stochastic comparisons for queueing models via random sums and intervals. Adv. Appl. Prob. 24, 960-985.

Joag-Dev, K. (1983) Independence via uncorrelatedness under certain dependence structures. Ann.Prob.11, 1037-1041.

Joag-Dev,K. and Proschan, F. (1983) Negative association of random variables with applications. Ann.Stat.11, 286-295.

Jogdeo, K. (1977) Association and probability inequalities. Ann.Stat.5, 495-504.

Jogdeo, K. (1978) On a probability bound of Marshall and Olkin. Ann.Stat.6, 232-234.

Kac, M. (1959) *Statistical Independence in Probability, Analysis and Number Theory,* Carus Math. Monogr. 12, Wiley, New York.

Kallenberg, O. (1983) *Random Measures*, Academic Press, London.

Kamae,T., Krengel,U. and O'Brien,G.C.(1977) Stochastic inequalities on partially ordered spaces. Ann.Prob.5,899-912.

Kanter, M.(1985) Lower bounds for the probability of overload in certain queueing networks. J.Appl.Prob.22,429-436.

Karlin, S. (1968) *Total Positivity*, Stanford University Press, Stanford, CA.

Karlin, S. and Novikoff, A. (1963) Generalized convex inequalities. Pacific J. Math. 13, 1251-1279.

Karlin, S. , Rinott, Y. (1980) Classes of orderings of measures and related correlation inequalities.I Multivariate totally positive distributions. J.Multiv.Anal.10,467-498.

Keilson, J. (1966) The ergodic queue length distribution for queueing systems with finite capacity. J. R. Statist. Soc. B 28, 190-201.

Keilson, J. (1978) Exponential spectra as a tool for the study of server-systems with several classes of customers. J. Appl. Prob. 15, 162-170.

Keilson, J. (1979) *Markov Chain Models-Rarity and Exponentiality*, Springer-Verlag, New York.

Kelbert,M.YA., Kontsevich,M.L., and Rybko,A.N.(1986) On Jackson networks on denumerable graphs. Manuscript.

Keilson, J. and Sumita, U. (1982) Uniform stochastic ordering and related inequalities. The Canadian Journal of Statistics 10, 181-198.

Kelley (1959) Measures on Boolean algebras. Pacific J. Math. 9, 1165-1176.

Kelly, F.(1979) *Reversibility and Stochastic Networks.* John Wiley, New York.

Kendall, D.G. (1953) Stochastic processes occurring in the theory of queues and their analysis by the method of the imbedded Markov chain. Ann. Math. Stat. 24, 338-354.

Khintchine, A. Ya. (1960) *Mathematical Methods in the Theory of Queueing*, Griffin, London.

Kijima, M. (1992) Further monotonicity properties of renewal processes. Adv. Appl. Prob. 25, 575-588.

Kingman, J.F.C. (1962) Some inequalities for the queue G/G/1. Biometrika 49, 315-324.

Kingman, J.F.C. (1972) *Regenerative Phenomena*, John Wiley & Sons, London.

Kőllerstrőm, J. (1976) Stochastic bounds for the single server queue. Math. Proc. Camb. Phil. Soc. 80, 521-525.

Kwiecinski, A. and Szekli, R.(1991) Compensator conditions for stochastic ordering of point processes. J.Appl.Prob.28, 751-761.

Kwiecinski, A. and Szekli, R. (1994) A class of associated point processes. Manuscript.

Langberg, N.A., Leon, R., Lynch, J. and Proschan, F. (1980) Extreme points of the class of discrete decreasing failure rate life distributions. Math. Oper. Res. 5, 35-42.

Langberg, N.A., Leon, R., Lynch, J. and Proschan, F. (1981) The extreme points of the set of decreasing failure rate distribution. Z. Wahrsch. Geb. 57, 303-310.

Last, G. (1993) On dependent marking and thinning of point processes. Stoch. Proc. Appl. 45, 73-94.

Last, G. and Brandt,A. (1994) *Marked Point Processes on the Real Line:The Dynamical Approach.* Lecture Notes, Springer .

Lebowitz, J.L. (1972) Bounds on the correlations and analyticity properties of ferromagnetic Ising spin systems. Commun.Math.Phys.28, 313-321.

Lehmann, E.L. (1966) Some concepts of dependence. Ann.Math.Stat. 37,1137-1153.

Leon, R. and Lynch,J. (1983) Preservation of life distribution classes under reliability operations. Math. Oper. Res. 8, 159-169.

Liggett,T.M.(1985) Interacting Particle Systems. Springer-Verlag, New York.

Lindley, D.V. (1952) The theory of queues with a single server. Proc. Camb. Phil. Soc. 48, 277-289.

Lindqvist,B.M.(1988) Association of probability measures on partially ordered spaces. J.Multiv.Anal.26,111-132.

Lindvall, T. (1986) On coupling of renewal processes with use of failure rates. Stoch. Proc. Appl. 22, 1-15.

Lindvall, T. (1988) Ergodicity and inequalities in a class of point processes. Stoch. Proc. Appl. 30, 121-131.

Lindvall, T. (1992) *Lectures on the Coupling Method.* Wiley, New York.

Loynes, R.M. (1962) The stability of a queue with non-independent interarrival and service times. Proc. Camb. Phil. Soc. 58, 497-520.

Lukacs, E. (1970) *Characteristic Functions*, 2nd Edition, Griffin, London.

Marshall, and A.W., Olkin, I. (1966) A multivariate exponential distribution. J.Amer. Stat. Assoc.62, 30-44.

Marshall, A.W. and Olkin, I. (1979) *Inequalities: Theory of Majorization and Applications.* Academic Press, New York.

Marshall, A.W. and Proschan, F. (1972) Classes of distributions applicable in replacement, with renewal theory implications. Proc. 6th Berkeley Symp. Math. Statist. Probab. 1970, Vol. 1, 495-515, Univ. California press, Berkeley.

Massey, W.A.(1987) Stochastic ordering for Markov processes on partially ordered spaces. Math.Oper.Res.12,350-367.

Massey, W.A.(1989) Stochastic ordering for birth-death-migration processes. Stoch. Proc. Appl.

Menich, R. and Serfozo, R.F. (1991) Optimality of routing and servicing in dependent parallel processing systems. Queueing Systems 9, 403-418.

Meester, L. and Shanthikumar, J.G. (1992) Regularity of stochastic processes: a theory based on directional convexity. P. E. I. S.

Meyn, S.P. and Tweedie, R.L. (1993a) *Markov Chains and Stochastic Stability.* Springer-Verlag, Berlin.

Meyn, S.P. and Tweedie, R.L. (1993b) Stability of Markovian processes II: continuous-time processes and sampled chains. Adv. Appl. Prob. 25, 487-517.

Miyazawa, M. (1989) Comparison of the loss probability of the $GI^X/GI/1/k$ queues with a common traffic intensity. J. Oper. Res. Soc. Japan 32, 505-516.

Miyazawa, M. (1990) Complementary generating functions for the $M^X/G/1/k$ and $GI/M^Y/1/k$ queues and their application to the comparison of loss probabilities. J.Appl. Prob. 27, 684-692.

Nachbin, L. (1965) *Topology and Order.* Van Nostrand, New York.

Newman, C.M. (1984) Asymptotic independence and limit theorems for positively and negatively dependent random variables. Inequalities in Statistics and Probability. IMS Lecture Notes Monograph Series 5,127-140.

Mc Nickle, D.(1991) Lagged queue-length correlations in two-node networks. Queueing Systems 8,97-104.

Niu, S.C.(1981) On the comparison of waiting times in tandem queues. J. Appl. Prob. 18, 707-714.

Niu, S.C. and Cooper, R.B. (1991) Transform-free analysis of M/G/1/K and related queues. Math. Oper. Res.

Norros, I. (1986) A compensator represenmtation of multivariate life length distributions, with applications. Scand. J. Statist. 13, 99-112.

Palm, C. (1943) Intensitätsschwankungen im Fernsprechverkehr. Ericsson Technics 44, 1-189.

Patuwo, B.E., Disney, R.L. and McNickle, D.C.(1991) The effect of correlated arrivals on queues. IIE Transactions.

Polya, G. and Szegö, G. (1925) *Aufgaben und Lehrsätze aus der Analysis*, V.I. Springer Verlag, Berlin.

Prabhu, N.U. (1965) *Queues and Inventories*, Wiley, New York.

Reich, E.(1957) Waiting times when queues are in tandem. Ann. Math. Stat. 28, 768-773.

Reuter, H. and Riedrich, T. (1981) On maximal sets of functions compatible with a partial ordering of distribution functions. Math. Operat.-forschung Stat. Ser. Optimiz. 12, 597-606.

Ridder,A.A.(1987) Stochastic Inequalities for Queues. Thesis.

Roberts, A.W. and Varberg, P.E. (1973) *Convex Functions*, Academic Press, New York.

Rolski, T. (1976) Order relations in the set of probability distribution functions and their applications in queueing theory. Dissertationes Math. 132.

Rolski, T. (1986) Upper bounds for single server queues with doubly stochastic Poisson arrivals. Math. Oper. Res. 11, 442-450.

Rolski, T. (1981) *Stationary Random Processes Associated with Point Processes*, Springer Lecture Notes Stat. 5. Springer Verlag, New York.

Rolski, T. and Stoyan, D. (1976) On the comparison of waiting times in GI/G/1 queues. Operations Research 24, 197-200.

Rolski, T. and Szekli, R. (1991) Stochastic ordering and thinning of point processes. Stoch. Proc. Appl. 37, 299-312.

Ross, S. (1983)*Stochastic Processes*. Wiley, New York.

Rüschendorf, L. (1980) Inequalities for the expectation of Δ monotone functions. Z. Wahrsch. Geb. 54, 341-349.

Rüschendorf, L. (1981) Ordering of distributions and rearrangement of functions. Ann. Prob. 9, 276-283.

Ryll-Nardzewski, C. (1961) Remarks on processes of calls. Proc. Fourth Berkeley Symp. Math. Statist. Probab. 2, 455-465.

Saaty, (1960) Opns. Res. 8, 755-772.

Schassberger, R. (1973) *Warteschlangen.* Springer-Verlag, Wien/New York.

Serfozo,R.(1989) Poisson functionals of Markov processes and queueing networks. Adv. Appl. Prob. 21, 595-611.

Serfozo,R.(1990) Reversibility and compound birth-death and migration processes. Preprint.

Serfozo, R. (1990) *Point Processes*, Chapter 1 in D.P. Heyman, M.J. Sobel, Eds., Handbooks in OR & MS, Vol. 2, Elsevier Science Publishers B.V., North-Holland.

Sevastyanov, B.A. (1957) An ergodic theorem for Markov processes and its application to telephone systems with refusals. Theory Prob. Appl. 2, 104-112.

Shaked, M. (1982) Dispersive ordering of distributions. J. Appl. Prob. 19, 310-320.

Shaked, M. (1982) A general theory of some positive dependence notions. J.Multiv. Anal.12,199-218.

Shaked, M. and Shanthikumar, J.G. (1987) The multivariate hazard construction. Stoch. Proc. Appl. 24, 241-258.

Shaked,M. and Shanthikumar, J.G. (1989) Multivariate stochastic orderings and positive dependence in reliability theory. Manuscript.

Shaked,M. and Shanthikumar, J.G. (1993) *Stochastic Orders and Their Applications.* Academic Press.

Shaked, M. and Szekli, R. (1993) Comparison of replacement policies via point processes. Manuscript.

Shaked, M. and Zhu, H. (1992) Some results on block replacement policies and renewal theory. J. Appl. Prob. 29, 932-946.

Shanbhag, D.N. and Tambouratzis, D.G. (1973) Erlang's formula and some results on the departure process for a loss system. J.Appl. Prob. 10, 233-240.

Shanthikumar, J.G. (1987) On stochastic comparison of random vectors. J. Appl. Prob. 24, 123-136.

Shanthikumar, J.G. (1988) DFR property of first passage times and its preservation under geometric compounding. Ann. Prob. 16,397-406.

Shanthikumar, J.G. and Yao, D. (1991) Strong stochastic convexity: closure properties and applications. J. Appl. Prob. 28, 131-145.

Snyder, D.L. and Miller, M.I. (1991) *Random Point Processes in Time and Space*, 2nd Edition, Springer-Verlag, New York.

Sobel, M. (1980) Simple inequalities for multiserver queues. Management Science 26, 951-956.

Stadje, W. (1985) The busy period of the queueing system $M/G/\infty$. J. Appl. Prob. 22, 697-704.

Stidham, S.Jr. (1974) A last word on $L=\lambda W$. Oper. Res. 22, 417-421.

Stoyan, D.(1983) *Comparison Methods for Queues and Other Stochastic Models.* John Wiley, Berlin.

Strassen, V. (1965) The existence of probability measures with given marginals. Ann. Math. Statist. 36, 423-439.

Sumita, U. and Shanthikumar, J.G. (1988) An age dependent counting process generated from a renewal process. Adv. Appl. Prob. 20, 739-755.

Szekli, R. (1986) On the concavity of the waiting time distribution in some GI/G/1 queues. J. Appl. Prob. 23, 555-561.

Szekli, R. (1987) The complete monotonicity of the waiting time density in some GI/G/k systems. Queueing Systems 1, 401-406.

Szekli, R. (1987) A note on moments inequalities for order statistics from starshaped distributions. Applicationes Mathematicae 19, 65-68.

Szekli, R. (1990) On the concavity of the infinitesimal renewal function. Stat. and Prob. Letters 10, 181-184.

Szekli, R., Disney,R.L. Hur.S. (1993a) On performance comparison of MR/GI/1 queues. Queueing Systems.

Szekli, R. Disney, R.L. Hur, S. (1993b) MR/GI/1 queues with positively correlated arrival stream. J. Appl. Prob. 31, 497-514.

Takacs, L. (1969) On Erlang's formula. Ann. Math. Stat. 40, 71-78.

Tchen, A. (1980) Inequalities for distributions with given marginals. Ann. Prob. 8. 811-827.

Tong, Y.L. (1980) *Probability Inequalities in Multivariate Distributions*, Academic Press, New York.

Van der Wal, J. (1989) Monotonicity of the throughput of a closed exponential queueing network in the number of jobs. OR Spektrum 11, 97-100.

Walrand, J. (1988) *An Introduction to Queueing Networks*, Prentice Hall, Englewood Cliffs.

Watanabe, S. (1964) On discontinuous additive functionals and Levy measures of a Markov process. Japanese J. Math. 34, 53-70.

Whitt, W. (1980) Uniform conditional stochastic order. J. Appl. Prob. 17, 112-123.

Whitt, W. (1981) Comparing counting processes and queues. Adv. Appl. Prob. 13, 207-220.

Widder, D.V. (1946) *The Laplace Transform*, Princeton University Press, Princeton.

Wolff, R.W. (1982) Poisson arrivals see time averages. Operations Res. 30, 223-231.

Index

Lecture Notes in Statistics

For information about Volumes 1 to 9
please contact Springer-Verlag

Vol. 10: A.A. McIntosh, Fitting Linear Models: An Application of Conjugate Gradient Algorithms. vi, 200 pages, 1982.

Vol. 11: D.F. Nicholls and B.G. Quinn, Random Coefficient Autoregressive Models: An Introduction. v, 154 pages, 1982.

Vol. 12: M. Jacobsen, Statistical Analysis of Counting Processes. vii, 226 pages, 1982.

Vol. 13: J. Pfanzagl (with the assistance of W. Wefelmeyer), Contributions to a General Asymptotic Statistical Theory. vii, 315 pages, 1982.

Vol. 14: GLIM 82: Proceedings of the International Conference on Generalised Linear Models. Edited by R. Gilchrist. v, 188 pages, 1982.

Vol. 15: K.R.W. Brewer and M. Hanif, Sampling with Unequal Probabilities. ix, 164 pages, 1983.

Vol. 16: Specifying Statistical Models: From Parametric to Non-Parametric, Using Bayesian or Non-Bayesian Approaches. Edited by J.P. Florens, M. Mouchart, J.P. Raoult, L. Simar, and A.F.M. Smith, xi, 204 pages, 1983.

Vol. 17: I.V. Basawa and D.J. Scott, Asymptotic Optimal Inference for Non-Ergodic Models. ix, 170 pages, 1983.

Vol. 18: W. Britton, Conjugate Duality and the Exponential Fourier Spectrum. v, 226 pages, 1983.

Vol. 19: L. Fernholz, von Mises Calculus For Statistical Functionals. viii, 124 pages, 1983.

Vol. 20: Mathematical Learning Models — Theory and Algorithms: Proceedings of a Conference. Edited by U. Herkenrath, D. Kalin, W. Vogel. xiv, 226 pages, 1983.

Vol. 21: H. Tong, Threshold Models in Non-linear Time Series Analysis. x, 323 pages, 1983.

Vol. 22: S. Johansen, Functional Relations, Random Coefficients and Nonlinear Regression with Application to Kinetic Data, viii, 126 pages, 1984.

Vol. 23: D.G. Saphire, Estimation of Victimization Prevalence Using Data from the National Crime Survey. v, 165 pages, 1984.

Vol. 24: T.S. Rao, M.M. Gabr, An Introduction to Bispectral Analysis and Bilinear Time Series Models. viii, 280 pages, 1984.

Vol. 25: Time Series Analysis of Irregularly Observed Data. Proceedings, 1983. Edited by E. Parzen. vii, 363 pages, 1984.

Vol. 26: Robust and Nonlinear Time Series Analysis. Proceedings, 1983. Edited by J. Franke, W. Härdle and D. Martin. ix, 286 pages, 1984.

Vol. 27: A. Janssen, H. Milbrodt, H. Strasser, Infinitely Divisible Statistical Experiments. vi, 163 pages, 1985.

Vol. 28: S. Amari, Differential-Geometrical Methods in Statistics. v, 290 pages, 1985.

Vol. 29: Statistics in Ornithology. Edited by B.J.T. Morgan and P.M. North. xxv, 418 pages, 1985.

Vol 30: J. Grandell, Stochastic Models of Air Pollutant Concentration. v, 110 pages, 1985.

Vol. 31: J. Pfanzagl, Asymptotic Expansions for General Statistical Models. vii, 505 pages, 1985.

Vol. 32: Generalized Linear Models. Proceedings, 1985. Edited by R. Gilchrist, B. Francis and J. Whittaker. vi, 178 pages, 1985.

Vol. 33: M. Csörgo, S. Csörgo, L. Horváth, An Asymptotic Theory for Empirical Reliability and Concentration Processes. v, 171 pages, 1986.

Vol. 34: D.E. Critchlow, Metric Methods for Analyzing Partially Ranked Data. x, 216 pages, 1985.

Vol. 35: Linear Statistical Inference. Proceedings, 1984. Edited by T. Calinski and W. Klonecki. vi, 318 pages, 1985.

Vol. 36: B. Matérn, Spatial Variation. Second Edition. 151 pages, 1986.

Vol. 37: Advances in Order Restricted Statistical Inference. Proceedings, 1985. Edited by R. Dykstra, T. Robertson and F.T. Wright. viii, 295 pages, 1986.

Vol. 38: Survey Research Designs: Towards a Better Understanding of Their Costs and Benefits. Edited by R.W. Pearson and R.F. Boruch. v, 129 pages, 1986.

Vol. 39: J.D. Malley, Optimal Unbiased Estimation of Variance Components. ix, 146 pages, 1986.

Vol. 40: H.R. Lerche, Boundary Crossing of Brownian Motion. v, 142 pages, 1986.

Vol. 41: F. Baccelli, P. Brémaud, Palm Probabilities and Stationary Queues. vii, 106 pages, 1987.

Vol. 42: S. Kullback, J.C. Keegel, J.H. Kullback, Topics in Statistical Information Theory. ix, 158 pages, 1987.

Vol. 43: B.C. Arnold, Majorization and the Lorenz Order: A Brief Introduction. vi, 122 pages, 1987.

Vol. 44: D.L. McLeish, Christopher G. Small, The Theory and Applications of Statistical Inference Functions. vi, 124 pages, 1987.

Vol. 45: J.K. Ghosh (Editor), Statistical Information and Likelihood. 384 pages, 1988.

Vol. 46: H.-G. Müller, Nonparametric Regression Analysis of Longitudinal Data. vi, 199 pages, 1988.

Vol. 47: A.J. Getson, F.C. Hsuan, {2}-Inverses and Their Statistical Application. viii, 110 pages, 1988.

Vol. 48: G.L. Bretthorst, Bayesian Spectrum Analysis and Parameter Estimation. xii, 209 pages, 1988.

Vol. 49: S.L. Lauritzen, Extremal Families and Systems of Sufficient Statistics. xv, 268 pages, 1988.

Vol. 50: O.E. Barndorff-Nielsen, Parametric Statistical Models and Likelihood. vii, 276 pages, 1988.

Vol. 51: J. Hüsler, R.-D. Reiss (Editors), Extreme Value Theory. Proceedings, 1987. x, 279 pages, 1989.

Vol. 52: P.K. Goel, T. Ramalingam, The Matching Methodology: Some Statistical Properties. viii, 152 pages, 1989.

Vol. 53: B.C. Arnold, N. Balakrishnan, Relations, Bounds and Approximations for Order Statistics. ix, 173 pages, 1989.

Vol. 54: K.R. Shah, B.K. Sinha, Theory of Optimal Designs. viii, 171 pages, 1989.

Vol. 55: L. McDonald, B. Manly, J. Lockwood, J. Logan (Editors), Estimation and Analysis of Insect Populations. Proceedings, 1988. xiv, 492 pages, 1989.

Vol. 56: J.K. Lindsey, The Analysis of Categorical Data Using GLIM. v, 168 pages, 1989.

Vol. 57: A. Decarli, B.J. Francis, R. Gilchrist, G.U.H. Seeber (Editors), Statistical Modelling. Proceedings, 1989. ix, 343 pages, 1989.

Vol. 58: O.E. Barndorff-Nielsen, P. Blæsild, P.S. Eriksen, Decomposition and Invariance of Measures, and Statistical Transformation Models. v, 147 pages, 1989.

Vol. 59: S. Gupta, R. Mukerjee, A Calculus for Factorial Arrangements. vi, 126 pages, 1989.

Vol. 60: L. Györfi, W. Härdle, P. Sarda, Ph. Vieu, Nonparametric Curve Estimation from Time Series. viii, 153 pages, 1989.

Vol. 61: J. Breckling, The Analysis of Directional Time Series: Applications to Wind Speed and Direction. viii, 238 pages, 1989.

Vol. 62: J.C. Akkerboom, Testing Problems with Linear or Angular Inequality Constraints. xii, 291 pages, 1990.

Vol. 63: J. Pfanzagl, Estimation in Semiparametric Models: Some Recent Developments. iv, 112 pages, 1990.

Vol. 64: S. Gabler, Minimax Solutions in Sampling from Finite Populations. v, 132 pages, 1990.

Vol. 65: A. Janssen, D.M. Mason, Non-Standard Rank Tests. vi, 252 pages, 1990.

Vol. 66: T. Wright, Exact Confidence Bounds when Sampling from Small Finite Universes. xvi, 431 pages, 1991.

Vol. 67: M.A. Tanner, Tools for Statistical Inference: Observed Data and Data Augmentation Methods. vi, 110 pages, 1991.

Vol. 68: M. Taniguchi, Higher Order Asymptotic Theory for Time Series Analysis. viii, 160 pages, 1991.

Vol. 69: N.J.D. Nagelkerke, Maximum Likelihood Estimation of Functional Relationships. v, 110 pages, 1992.

Vol. 70: K. Iida, Studies on the Optimal Search Plan. viii, 130 pages, 1992.

Vol. 71: E.M.R.A. Engel, A Road to Randomness in Physical Systems. ix, 155 pages, 1992.

Vol. 72: J.K. Lindsey, The Analysis of Stochastic Processes using GLIM. vi, 294 pages, 1992.

Vol. 73: B.C. Arnold, E. Castillo, J.-M. Sarabia, Conditionally Specified Distributions. xiii, 151 pages, 1992.

Vol. 74: P. Barone, A. Frigessi, M. Piccioni (Editors), Stochastic Models, Statistical Methods, and Algorithms in Image Analysis. vi, 258 pages, 1992.

Vol. 75: P.K. Goel, N.S. Iyengar (Editors), Bayesian Analysis in Statistics and Econometrics. xi, 410 pages, 1992.

Vol. 76: L. Bondesson, Generalized Gamma Convolutions and Related Classes of Distributions and Densities. viii, 173 pages, 1992.

Vol. 77: E. Mammen, When Does Bootstrap Work? Asymptotic Results and Simulations.vi, 196 pages, 1992.

Vol. 78: L. Fahrmeir, B. Francis, R. Gilchrist, G. Tutz (Editors), Advances in GLIM and Statistical Modelling: Proceedings of the GLIM92 Conference and the 7th International Workshop on Statistical Modelling, Munich, 13-17 July 1992. ix, 225 pages, 1992.

Vol. 79: N. Schmitz, Optimal Sequentially Planned Decision Procedures. xii, 209 pages, 1992.

Vol. 80: M. Fligner, J. Verducci (Editors), Probability Models and Statistical Analyses for Ranking Data. xxii, 306 pages, 1992.

Vol. 81: P. Spirtes, C. Glymour, R. Scheines, Causation, Prediction, and Search. xxiii, 526 pages, 1993.

Vol. 82: A. Korostelev and A. Tsybakov, Minimax Theory of Image Reconstruction. xii, 268 pages, 1993.

Vol. 83: C. Gatsonis, J. Hodges, R. Kass, N. Singpurwalla (Editors), Case Studies in Bayesian Statistics. xii, 437 pages, 1993.

Vol. 84: S. Yamada, Pivotal Measures in Statistical Experiments and Sufficiency. vii, 129 pages, 1994.

Vol. 85: P. Doukhan, Mixing: Properties and Examples. xi, 142 pages, 1994.

Vol. 86: W. Vach, Logistic Regression with Missing Values in the Covariates. xi, 139 pages, 1994.

Vol. 87: J. Møller, Lectures on Random Voronoi Tessellations.vii, 134 pages, 1994.

Vol. 88: J.E. Kolassa, Series Approximation Methods in Statistics.viii, 150 pages, 1994.

Vol. 89: P. Cheeseman and R.W. Oldford (Editors), Selecting Models From Data: Artificial Intelligence and Statistics IV. x, 487 pages, 1994.

Vol. 90: A. Csenki, Dependability for Systems with a Partitioned State Space: Markov and Semi-Markov Theory and Computational Implementation. x, 241 pages, 1994.

Vol. 91: J.D. Malley, Statistical Applications of Jordan Algebras. viii, 101 pages, 1994.

Vol. 92: M. Eerola, Probabilistic Causality in Longitudinal Studies. vii, 133 pages, 1994.

Vol. 93: Bernard Van Cutsem (Editor), Classification and Dissimilarity Analysis. xiv, 238 pages, 1994.

Vol. 94: Jane F. Gentleman and G.A. Whitmore (Editors), Case Studies in Data Analysis. viii, 262 pages, 1994.

Vol. 95: Shelemyahu Zacks, Stochastic Visibility in Random Fields. x, 175 pages, 1994.

Vol. 96: Ibrahim Rahimov, Random Sums and Branching Stochastic Processes. viii, 195 pages, 1995.

Vol. 97: R. Szekli, Stochastic Ordering and Dependence in Applied Probablility, viii, 194 pages, 1995.

General Remarks

Lecture Notes are printed by photo-offset from the master-copy delivered in camera-ready form by the authors of monographs, resp. editors of proceedings volumes. For this purpose Springer-Verlag provides technical instructions for the preparation of manuscripts. Volume editors are requested to distribute these to all contributing authors of proceedings volumes. Some homogeneity in the presentation of the contributions in a multi-author volume is desirable.

Careful preparation of manuscripts will help keep production time short and ensure a satisfactory appearance of the finished book. The actual production of a Lecture Notes volume normally takes approximately 8 weeks.

For monograph manuscripts typed or typeset according to our instructions, Springer-Verlag can, if necessary, contribute towards the preparation costs at a fixed rate.

Authors of monographs receive 50 free copies of their book. Editors of proceedings volumes similarly receive 50 copies of the book and are responsible for redistributing these to authors etc. at their discretion. No reprints of individual contributions can be supplied. No royalty is paid on Lecture Notes volumes.

Volume authors and editors are entitled to purchase further copies of their book for their personal use at a discount of 33.3% and other Springer mathematics books at a discount of 20% directly from Springer-Verlag. Authors contributing to proceedings volumes may purchase the volume in which their article appears at a discount of 20%.

Springer-Verlag secures the copyright for each volume.

Series Editors:

Professor S. Fienberg
Department of Statistics
Carnegie Mellon University
Pittsburgh, Pennsylvania 15213
USA

Professor J. Gani
Stochastic Analysis Group, SMS
Australian National University
Canberra ACT 2601
Australia

Professor K. Krickeberg
3 Rue de L'Estrapade
75005 Paris
France

Professor I. Olkin
Department of Statistics
Stanford University
Stanford, California 94305
USA

Professor N. Wermuth
Department of Psychology
Johannes Gutenberg University
Postfach 3980
D-6500 Mainz
Germany